Expandability of the 21st Century Army

James A. Dewar

Steven C. Bankes

Sean J. A. Edwards

James C. Wendt

T0159548

Prepared for the
United States Army

Arroyo Center

RAND

For more information on the RAND Arroyo Center, contact the Director of Operations, (310) 393-0411, extension 6500, or visit the Arroyo Center's Web site at http://www.rand.org/organization/ard/

This project began during fiscal year 1997 at the request of COL Tom Molino and LTC Tim Daniel of DAMO-SSP. They asked that we develop a framework for thinking about expandability of the Army into the next 15–20 years. Their concern was that the Army, in the post–Cold War world and with the military drawdown, was not paying enough attention to the possibility that it might have to expand in the coming 15–20 years. They were further concerned that the drawdown and BRAC (base realignment and closure) processes might inadvertently damage the Army's ability to expand in the future.

We developed the framework and applied it to the current Army situation to see whether it suggested any near-term actions to preserve important expandability capabilities for the future. We recommended follow-on work to look at costs (to the extent possible) and, under the aegis of COL Richard Olson (DAMO-SSP), carried out that investigation during fiscal year 1998.

This report should be of interest to strategic planners in general and to Army strategic planners in particular. The results of the study draw broad outlines about the Army's expandability options now and into the future. The methods used to arrive at those results highlight a new capability to do exploratory modeling and should be of interest to any planner whose planning problem can be modeled parametrically.

This research was sponsored by Deputy Chief of Staff for Operations and Plans (DCSOPS), U.S. Army, and was conducted in RAND Arroyo Center's Strategy, Doctrine, and Resources Program. The Arroyo

Center is a federally funded research and development center sponsored by the United States Army.

CONTENTS

Preface   iii

Figures   ix

Tables   xiii

Summary   xv

Acknowledgments   xxix

Abbreviations   xxxi

Chapter One
    INTRODUCTION   1
    Why Study the Expandability of the 21st Century Army? . .   1
    Study Objectives   5
        Addressing the Issues   5
    Defining Expandability   6
    The Current State of Expandability Study   8
    Approach   10
        Current Versus Future Expansion   11
        Heavy Versus Light Expansion   12
        Expansion Timing Versus Expansion Timing and Cost .   13
    Report Outline   15

Chapter Two
    A FRAMEWORK FOR THINKING
    ABOUT EXPANDABILITY   17
    Framework Desiderata   17
    The Basics of Expandability   18
        Defining Capabilities   19

An Exploratory Framework   20
  Parametric Computer Models   21
  The Exploratory Modeling Environment   21
  Documenting Exploratory Modeling Outcomes   25
Addressing Expandability Issues   25
  Description/Exploration Versus Prescription   25
  Developing Capabilities over Time   26
  Expandability Issues   27
Summary   30

Chapter Three
EXPANDING THE ARMY'S HEAVY DIVISIONS TODAY ...   31
Trainees   33
  Active Component or "Ready" Brigades   33
  Enhanced Ready Brigades   33
  ARNG Brigades   34
  New Units   35
Training Processes, Facilities, and Personnel   38
  Advanced Training   38
  Training Up for ARNG Deficiencies   43
  New Unit Training   45
Equipment Production   50
  Division Sets Required   50
  Expansion Capability   58
Maximum Expansion Rates Today   67
  Expanding Personnel   67
  Expanding Equipment   69
  Maximum Expansion Rates   70

Chapter Four
EXPANDING THE ARMY'S HEAVY BRIGADES
IN THE FUTURE   73
The Model   73
Documenting an Exploratory Analysis   74
Bottlenecks in Expanding Heavy Brigades   74
  One NTC Only   74
  Two NTCs   76
  Three NTCs   78
  More Than Three NTCs   80
Summary   82

Chapter Five
    COSTS AND EXPANDING THE ARMY'S HEAVY FORCES
    IN THE FUTURE 85
    Introduction 85
    The Cost of Expanding Heavy Forces Equipment 87
    Definitions 89
        Recurring Costs 89
        Expansion Costs 90
    The Cost Model and Parameters 90
    Recurring Costs Versus the Time to Expand Heavy
        Forces 91
    Recurring and Expansion Costs 95
        Three NTCs 95
        Six NTCs 101
    Conclusions 104
    Worry Curves 105

Chapter Six
    EXPANDING THE ARMY'S LIGHT DIVISIONS TODAY . . . 107
    Expansion Process. 108
    Training Processes, Facilities, and Personnel 110
        Advanced Training 111
        Training Up for ARNG Deficiencies 116
        New Unit Training 116
    Trainees 118
        Active Component or "Ready" Brigades 118
        ARNG Brigades 118
        New Units 119
    Maximum Production Rate 121

Chapter Seven
    EXPANDING THE ARMY'S LIGHT BRIGADES
    IN THE FUTURE 123
    The Model 123
    Bottlenecks in Expanding Light Forces 124
        One JRTC Only. 124
        Two JRTCs 126
        Three or More JRTCs 128
    Summary 130

Chapter Eight
  COSTS AND EXPANDING THE ARMY'S LIGHT FORCES
  IN THE FUTURE   133
  Introduction   133
  Recurring Costs Versus the Time to Expand Light Forces .   134
  Recurring and Expansion Costs   136
    Three JRTCs.   137
    Six JRTCs   141
  Conclusions   144

Chapter Nine
  CONCLUSIONS.   147
  Heavy Force Expansion Timing   147
  Heavy Force Expansion Timing and Cost   149
    Recurring Costs   149
    Recurring and Expansion Costs   150
  Light Force Expansion Timing   151
  Light Force Expansion Timing and Cost   151
    Recurring Costs   152
    Recurring and Expansion Costs   152
  General Conclusions   153
  Recommendations   154
    Things to Do   154
    Things to Watch For   155

Appendix

A. EXPLORATORY MODELING   157

B. EXPLORATORY MODELING PARAMETER VALUES   161

References   169

FIGURES

S.1. A Worry Curve xxvii
1.1. Army Troop Strength Since 1810 4
2.1. Sample Graph from the Exploratory Modeling Environment 23
2.2. Sample Set of Slider Bars 24
2.3. Sample Colored Region Plot 25
3.1. Flow Model 32
3.2. Advanced Training Cumulative Capacity 43
3.3. World War II Personnel Expansion Rate 46
3.4. OCS and BIOC Graduation of Infantry Officers 48
3.5. Abrams Production 52
3.6. Bradley Production 53
3.7. Apache Production 54
3.8. MLRS Production 55
3.9. Patriot Production 56
3.10. Maximum Abrams Production 62
3.11. Maximum Bradley Production 64
3.12. Maximum Apache Production 65
3.13. Maximum MLRS Production 66
3.14. Maximum Patriot Production 67
3.15. Advanced Training Capacity Versus Unit Availability . 68
3.16. Maximum Production: One Plant 69
3.17. Maximum Production: Two Plants 70
3.18. Maximum Personnel and Equipment Expansion 71
4.1. Flow of Parametric Model 73
4.2. Expanding Heavy Forces with One NTC 75
4.3. Expanding Heavy Forces with Two NTCs 77
4.4. Expanding Heavy Forces with Three NTCs 79

4.5. Expanding Heavy Forces with Four NTCs   81
5.1. Recurring Costs as a Function of the Number of
      Target Brigades and Target Month   92
5.2. Recurring Costs as a Function of the Number of NTCs
      and Target Month   93
5.3. Number of Initial Ready Brigades as a Function of the
      Target Brigades and Target Month   94
5.4. Minimal Recurring Costs with No More Than Three
      Allowable NTCs   96
5.5. Number of Initial Ready Brigades for Each of the
      Minimal-Recurring-Cost Force Structures   97
5.6. Expanding Heavy Forces with Three NTCs   97
5.7. Expansion Costs for Each of the Minimal-Recurring-
      Cost Force Structures   98
5.8. Initial Reserve Brigades for Each of the Minimal-
      Recurring-Cost Force Structures   99
5.9. Number of NTCs for Each of the Minimal-Recurring-
      Cost Force Structures   100
5.10. Minimal Recurring Costs with No More Than Six
      Allowable NTCs   101
5.11. Number of Initial Ready Brigades for Each of the
      Minimal-Recurring-Cost Force Structures   102
5.12. Expansion Costs for Each of the Minimal-Recurring-
      Cost Force Structures   103
5.13. Number of NTCs for Each of the Minimal-Recurring-
      Cost Force Structures   103
5.14. A Worry Curve   106
6.1. Light Force Expandability Flow Model   109
6.2. Advanced Light Forces Training Cumulative
      Capacity   116
6.3. JRTC Capacity Versus Unit Availability   121
7.1. Flow of Parametric Model   123
7.2. Expanding Light Forces with One JRTC   125
7.3. Effect on Expansion of Light Forces of Adding
      Advanced Training Sites   127
7.4. Effect on a Nominal Two-JRTC Expansion Capability
      (with Marginal Individual Training Capacity) of
      Doubling Other Delays   127

7.5.   Effect on a Nominal Three-JRTC Expansion Capability
        (with Marginal Individual Training Capacity) of
        Doubling Other Delays   129
7.6.   Effect on a Nominal Four-JRTC Expansion Capability
        (with Marginal Individual Training Capacity) of
        Doubling Other Delays   129
8.1.   Recurring Costs as a Function of the Number of
        Target Brigades and Target Month   135
8.2.   Recurring Costs as a Function of the Number of JRTCs
        and Target Month   136
8.3.   Minimal-Recurring-Costs with No More Than Three
        Allowable JRTCs   137
8.4.   Number of Initial Ready Brigades for Each of the
        Minimal-Recurring-Cost Force Structures   138
8.5.   Expansion Costs for the Minimal-Recurring-Cost
        Force Structures   138
8.6.   Initial Reserve Brigades for Each of the Minimal-
        Recurring-Cost Force Structures   140
8.7.   Number of JRTCs for Each of the Minimal-Recurring-
        Cost Force Structures   140
8.8.   Minimal-Recurring-Costs with No More Than Six
        Allowable JRTCs   142
8.9.   Number of Initial Ready Brigades for Each of the
        Minimal-Recurring-Cost Force Structures   142
8.10.   Expansion Costs for Each of the Minimal-Recurring-
        Cost Force Structures   143
8.11.   Number of JRTCs for Each of the Minimal-Recurring-
        Cost Force Structures   143
A.1.   The Central Challenge of Exploratory Modeling   158

TABLES

3.1. NCO Requirements 36
3.2. Advanced Training Activities 39
3.3. Nominal HDE Equipment 58
3.4. Production Summary 61
3.5. Manufacturing Ramp Rates 61
6.1. OPSGROUP Instructors 115
B.1. Heavy Force Timing Parameters 162
B.2. Heavy Force Cost Parameters 163
B.3. Light Force Timing Parameters 167
B.4. Light Force Cost Parameters 168

The Army actively studies and plans for near-term mobilization (the readying of reserves for action). Reconstitution—or adding new units to the Army—hasn't been studied seriously since then-President George Bush made it an important element in his post–Cold War defense policy in 1992. Current thinking, as expressed in the recent *Quadrennial Defense Review*, is that a global peer competitor—the primary reason for thinking about reconstitution of the Army—is unlikely to emerge in the period between now and 2015. If expandability is thought of as any increase in the capabilities of the Army beyond its current ready forces, there are at least three reasons for considering all expandability options in today's strategic planning out to 2015.

The first reason is the serious uncertainty we face in looking to the future. More so than during the Cold War, we don't know who the enemy is or will be, where it might emerge, what capabilities it may have (particularly if the so-called revolution in military affairs comes about), or how great the threat to the United States will be. This argues for a certain amount of hedging against the possibility that we are wrong about the size of the threat 20 years in the future.

Second, there is serious uncertainty about the future size of the Army. Any further shrinkage in the Army would make it that much easier for a competitor to become a peer and that much more important to consider options for eventually growing larger—if only to meet a larger number of lesser threats.

The third reason stems from the fact that the military, in general, has historically done a poor job of expanding in a timely manner and the fact that a peer competitor, historically, has always emerged eventually. These facts suggest that some thought about expansion would be prudent in a strategic planning exercise.

A fourth reason for looking anew at expandability stems from advances in computer technology and modeling capabilities that permit a more robust exploration of future scenarios.

The goals of this research were to:

- Develop a framework for studying expandability issues today and into the future, and
- Use that framework to explore whether plausible futures suggest action today to facilitate expansion capabilities.

The framework consists primarily of a parametric model and an exploratory modeling environment. Exploratory modeling was developed by one of the authors (Bankes). In its application for this research, it can be thought of as an automated "what if?" capability. Akin to a standard sensitivity analysis tool, the modeling environment allows, for example, graphical output to be examined interactively while the user moves slider bars that represent the model parameters and their acceptable ranges. This interactive capability permits a real-time directed tour through the model's outcome space looking for interesting areas that represent potential changes in the way expandability should be handled because of changed conditions in the future.

The model itself is a parametric bookkeeping model that keeps track of the major elements that go into an Army expansion. Both the parameters for that model and the ranges over which they could vary derived from extensive conversations with subject area experts.

The analytic approach divided the problem along three dimensions:

- Current versus future expansion.
- Heavy versus light expansion.
- Expansion timing only, versus expansion timing and cost.

In the first dimension, we took a close look at current expandability capabilities and constraints in order to develop simplified parametric models of expandability that could be exercised in the exploratory modeling environment to explore the future. The second dimension distinguished between expansions that significantly involved the industrial base and those that didn't. There are other differences between the two cases, to be sure, but the primary one has to do with whether or not a great deal of heavy equipment is involved. Since the production of heavy equipment is classically the slowest and most expensive part of any military expansion, looking at both pure heavy and pure light expansions provides a good feel for the slowest and fastest plausible expansions.

The third dimension was designed to pay attention to the fact that expandability has two distinct phases, with differing primary concerns. If there is an expansion taking place, the primary concern lies in how fast it can be accomplished—time is the most important factor. During peacetime, on the other hand, cost is a much more important factor. The recurring costs of an expansion capability dominate thinking, and the timing or cost of an eventual expansion recedes into the background. The research paid attention to both phases of expansion.

There were thus four major explorations—heavy and light expansion timing explorations, and heavy and light costing explorations. The results of the modeling explorations for each of these cases follow. The implications of these explorations for expandability are discussed in the "general conclusions" section.

EXPANSION TIMING

The primary means for exploring expansion timing in the future was the parametric model embedded in an exploratory modeling environment. With each aspect of the expansion system allowed to vary within a range of plausible future values, the exploratory modeling environment permitted a real-time, interactive exploration of the effects of changing a parameter value. By moving slider bars representing the parameters of the model, we explored a wide variety of different capacities, durations, and initial conditions in the expansion system to see under what conditions the system changed in important ways.

Heavy Forces

For heavy force expansions today, the bottleneck is training, particularly advanced brigade and division-level training. After there are sufficient trained brigades to man current equipment, the main impediment would become the ability of the industrial base to produce more equipment.

Explorations suggest that advanced training would remain the primary bottleneck to expansion under a wide variety of plausible future conditions. The most sensitive parameter in the model was the number of NTCs (combat training centers for heavy brigades). Using just one NTC (corresponding to the current National Training Center at Fort Irwin) during an expansion created a bottleneck so severe that the Army would basically need to have its capabilities in the ready force in order to meet a large threat. This remained true in the model not only under wide variations in training times and pre-expansion force structures, but also under all but the most severe industrial base variations.

Some other highlights of the timing explorations for heavy forces were the following:

- If ready troops required retraining, as soon as the retraining began, the system was unable to produce additional ready brigades unless additional NTCs were brought on line.

- In expansions beyond current reserve forces, accessions had to keep individual training sites full to keep up with three NTCs and had to be expanded to keep up with four or more. Said another way, recruiting and individual training could become bottlenecks in a large expansion in the model.

- Beyond the current inventory of 26 divisions of heavy equipment, the industrial base could become a bottleneck. One plant at full capacity could keep up with the manpower output of two NTCs, while two plants at full capacity could keep up with the output of four NTCs.

Light Forces

Without the requirement for a significant complement of heavy equipment,[1] the expansion timing of light forces will be driven primarily by advanced training at a facility like the Joint Readiness Training Center. In contrast to heavy force expansions, the time required to complete JRTC training is short enough that other factors could come into play more easily in the future.

Highlights of the modeled time explorations for light forces include the following:

- If only one JRTC was used for advanced training, its throughput would be the expansion bottleneck under all but extreme circumstances.

- The number of JRTCs was again a primary driver of decreased expansion time. With two or more JRTCs, several other factors could become bottlenecks. The most sensitive were the capacity of the individual training system and the advanced training time.

- Expansions beyond current or future reserve forces required an increase in recruitment or conscription and individual training sites for each increase in advanced training sites in order to keep recruiting and individual training from becoming bottlenecks.

- Retraining needs could, again, bring to a halt the production of new ready light units (unless the number of JRTCs was expanded further).

EXPANSION TIMING AND COST

In the timing analysis we were interested in how fast the Army could expand under a variety of conditions. For the costing explorations we chose a different focus. Because cost is most important during peacetime and time is most important during an expansion, the primary tradeoff of interest is between recurring (peacetime) costs of expansion capabilities and the time required to expand starting with those capabilities. As a way of standardizing the tradeoff between

---

[1]The one possible industrial base problem here was trucks. This point is discussed in footnote 24 of Chapter Three.

peacetime costs and expansion times, we looked at the costs of expanding to a given size in a given time. For that to be meaningful, we allowed the specification of all but the number of initial ready units. The model would then compute how much that system could expand in the given time and then require enough initial ready brigades in order to reach the target size. This ensures meeting of the size and time goals, and the resulting cost computation is the recurring costs of a system that could reach the target size in the target time with all but the number of initial ready troops arbitrarily selected.

A secondary costing concern is the costs during an expansion itself (though they are more likely to be a concern during a small expansion than during a large one). This led to two distinct explorations—one looking just at recurring costs versus time to expand, and the other looking at recurring and expansion costs.

Heavy Forces

Recurring costs. In looking at recurring costs, we were primarily interested in changes in recurring costs. For that reason we included only the recurring costs of the units and facilities that would be used during an expansion. This included the recurring costs of those ready and reserve units that would actually be trained during the expansion plus the recurring costs of the facilities (such as the NTC) that would be used and were being maintained before the expansion began. It also included the recurring costs of ready units that would be needed to man any additional training sites brought on line after the expansion began. Since the primary objective was comparative costs under a variety of conditions, the industrial base—which is likely to remain in a constant "warm" condition prior to an expansion—was dropped as a recurring cost.

There are no particular surprises in the costing explorations.

- For short expansion times (less than a year or so), most of the Army's required capabilities had to be in high readiness with a concomitant increase in recurring costs.

- For expansions that require fewer than six or so brigades per year, the current ability to expand was adequate and only reserve or civilian units were required to meet expansion timelines.

- As long as the recurring costs of a ready unit were substantially greater than those of reserve or civilian units, the number of initial ready brigades required to meet expansion size and time targets drove recurring costs.

- Expansion times between about 6 and 30 months saw the greatest benefit in reduced recurring costs when additional NTCs are brought on line during an expansion.

Recurring and expansion costs. If one were to know in any expansion exactly how large the Army should become and how long it had to reach that size, one could compute the minimum-recurring-cost force structure required to accomplish that expansion. Computing that minimum-recurring-cost force structure for a variety of size-time combinations provides some insight into the expansion costs and parameters associated with each such expansion. In this case, expansion costs are basically those costs during expansion that wouldn't have occurred if the expansion hadn't taken place. This includes the incremental costs of units that have been brought to readiness as well as the capital and operating costs of bringing new training facilities on line during the expansion.

Highlights of the modeling explorations for these minimum-recurring-cost structures are as follows:

- Minimum recurring costs varied smoothly over size-time space.

- Expansion costs exhibited two competing characteristics: (1) the general increase in costs due solely to the increased time that trained-up reserve and civilian units had to be maintained, and (2) a slow decrease in costs due to a decreased need over time to maintain reserve units or bring new NTCs on line in order to meet size and time targets.

- Where recurring costs were highest, expansion costs tend to be lowest (as there basically *was* no expansion possible because of the short timelines).

- Long, small expansions required the lowest recurring and expansion costs.

- Long, large expansions incurred large expansion costs but low recurring costs.

- Large expansions in the (roughly) 12- to 30-month time range were the most costly.

Light Forces

As with heavy forces, there were no particular surprises in the costing explorations. In the light forces case, however, expansion could take place more quickly.

Recurring costs

- Recurring costs for short (in this case, half-a-year or less) expansion times were high and in direct proportion to the target expansion size.

- Expansions as fast as four divisions per year could be accomplished with today's facilities and only reserve and civilian brigades would be needed to meet the timelines.

- Expansion timelines between (roughly) 6 and 24 months saw the greatest benefit in reduced recurring costs when additional JRTCs were brought on line during expansion.

Recurring and expansion costs

- Minimum recurring costs varied smoothly (though more quickly than for heavy force expansion) over size-time space and were driven primarily by the number of initial ready brigades required to meet size and time targets.

- Expansion costs exhibited the same two competing characteristics except in the light forces case; these competing characteristics produced a more varied picture across size-time space.

- As in the heavy case, where recurring costs were highest, expansion costs tend to be lowest, and long, small expansions required the lowest recurring and expansion costs.

- Long, large expansions incurred large expansion costs, but low recurring costs and large expansions in the (roughly) 6- to 24-month time range were most costly.

GENERAL CONCLUSIONS

The main conclusion one can draw from the explorations with the parametric model is that generally only extreme values of the parameters upset one's intuition about the nature or cause of bottlenecks in the modeled system. How likely is that to be true for "real" expansions?

The parametric model used was primarily a bookkeeping model of an expansion pipeline system with parameters for the dwell times at various stages of the pipeline. The model was verified in the sense that it kept the books properly. It was not validated in the sense that it would predict future expansion times. On the other hand, if any given set of modeled dwell times were to obtain for a real expansion, the total modeled expansion time would, of course, be the actual expansion time. By ranging across a robust set of plausible timing values, then, the explorations with the model cover most of the plausible delays in any expansion of the real Army. For that reason, general conclusions about the model explorations transfer well to reality. Beyond that, there are a few additional conclusions that can be drawn.

Expansion depends primarily on bringing people into the Army, training them, and equipping them. There are two general characteristics one would wish for an expansion system. The most important characteristic is that one would want the "pipes" in the expansion system to be "short" and "wide" enough to produce an expanded Army in time. The shorter the time allowed, the less able to comply will any expansion system be, and the more must the Army rely on ready units. In today's Army the expansion pipeline is quite "wide" in places because of large reserves of both trained soldiers and equipment. This means the primary constraint on expansion today (and into many plausible futures) is training—specifically advanced training at the brigade and division levels. If training could be shortened (e.g., by reducing the training time required) and/or widened (e.g., by increasing the number of training sites), today's expansion capability could be improved. Either seems unlikely in today's climate—the first because of an increase in the complexity of modern operations, the need for more joint and coalition training, and the potential for information operations to add an additional dimension to training, and the second because of declining military budgets. Further, the characteristics of today's system are likely to degrade with time (although if the equipment does not become militarily obsolete, it will remain functional throughout the 15–20 year period of study).

The other desirable—but less important—characteristic in an expansion system is that it be balanced, in the sense that there are no serious bottlenecks. The clear imbalance today in the heavy force expansion system is training, but that is due to the fact that the Army maintains a large reserve structure ready to be trained (which masks any imbalance in the recruiting and individual training part of the pipeline) and due to the Reagan buildup of equipment (which masks any imbalance in the industrial base part of the system). The heavy expansion system would require more NTCs in order to keep up with the steady-state production out of the Army's individual training sites. For expansions beyond existing equipment levels (currently about 26 divisions of heavy equipment), the expansion system would require an industrial base with something like two plants operating two and a half work shifts (the practical maximum) to stay in balance with the remainder of the system.

The light force expansion system, by contrast, is well balanced. As fast as people can be brought into the expansion pipeline, they can be moved through it and out the other end. Any increase in the training capability must be accompanied by an increase in recruiting and individual training in order to keep the light force expansion pipeline balanced.

RECOMMENDATIONS

The need to expand today's Army is small. The perceived need to be able to expand the Army any time soon is also small. Even so, there are low-cost actions that would enhance the Army's ability to expand in the future that are suggested by this analysis. These include both things to do and things to watch for.

Things to Do

- The biggest positive effect on either heavy or light expansion capabilities would be to decrease the advanced training time. If this can be done without decreasing military capability, it is the easiest means for decreasing expansion timelines both today and into most plausible futures.

- In the event of a sizable heavy expansion, it would take approximately three NTCs to provide a balanced expansion capability both with today's (and most future) individual training capabilities and capacities as well as equipment stocks. Both heavy and light expansions of a half-year or longer would benefit from further combat training centers (CTCs). The Army would be well served to prepare and maintain plans for building additional CTCs (and individual training sites if required). This should particularly include where and how they would be built and who would man them.[2]

- Whether or not there is a coming revolution in military affairs, the best preparation for a large future expansion would be for the Army to maintain and upgrade its current large equipment inventory. As the National Defense Panel's report on transforming defense put it, "It is more important to have a weapon on hand in adequate quantities than to have the capability available to produce that weapon six months or a year later."[3]

- In any large expansion, recruiting could become a bottleneck to the expansion system. A clear and historically employed solution for this is to conscript people into the Army. That remains a viable solution and it behooves today's all-volunteer Army to retain the ability to implement a conscription system on short notice.

---

[2]If heavy required forces were to be deployed in or near Europe, the current facilities at Hohenfelds in Germany could be used. For light forces, Fort Chaffey—the original JRTC site—could most easily and quickly be developed into an additional advanced training site.

[3]*Transforming Defense: National Security in the 21st Century*, Report of the National Defense Panel, December 1997.

Things to Watch For

Many of the things to watch for are generally self-evident and currently being done. Nonetheless, it is useful to detail them as a means of reinforcing their connection with the Army's ability to expand in the future.

- Any developments that would make obsolete our current equipment would seriously degrade our expandability capabilities (if not our then-current forces). Such an occurrence would signal the need for a reevaluation of our industrial base policy (perhaps suggesting a change in its readiness status). Any serious decline in the functionality of the current equipment stocks would also degrade expandability capability, bringing the much "longer" and "narrower" industrial base into play.

- Any significant changes in training time (positive or negative) at the NTC or JRTC would affect expandability capabilities (negatively and positively, respectively). Training times should be adequate for proper preparation of Army troops, but changes in those times need to be factored into the Army's expansion capabilities, in case changes need to be made to those capabilities.

- The Army should look for any hint of a need to expand in size. Any expansion, particularly in tight budgetary times, is likely to be dominated more by political than military concerns. If the Army is ever to expand again, it should begin making the case for expansion as soon as arguable indications of that need arise.

- The Army generally does a good job of recruiting for its manpower needs. It also pays close attention to changes in the willingness in U.S. society to serve in the Army. That should continue and be connected into thinking about expansion capabilities. Any expansion that goes beyond then-current reserve forces will be directly dependent on the willingness of young people to serve.

- The final signpost is the most obvious. Because of the critical dependency of having ready troops for military situations that provide short warning time, the Army should continue to monitor for threats that would stress its current ready capabilities. There are those who would argue that this is already the case.

Indeed, the ready military capabilities required to ensure national safety will always be debatable. This study has given the Army a framework and preliminary guidelines for addressing questions of when, how, and by how much the Army should expand to meet demands that exceed its then-current capabilities now and into the future.

"WORRY CURVES"

There is one additional extra-analytic capability provided by the exploratory modeling environment that is worth mentioning. Consider the following: When contemplating how big the Army should be and how quickly it ought to be able to expand, a military analyst or planner is thinking about a curve in expansion size-time space that describes roughly how big the Army should be able to get as a function of warning time in order to carry out its security mandate. Figure S.1 is a depiction of such a "worry curve." In words, the worry curve says that the Army should have about 30 ready brigades (10 divisions) in order to handle immediate crises and be able to

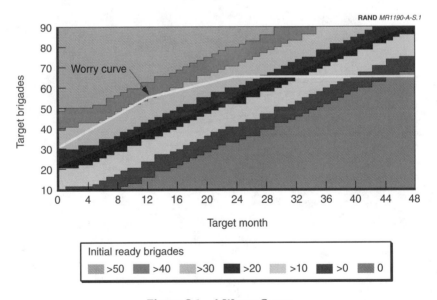

Figure S.1—A Worry Curve

ramp up to as many as 69 ready brigades (23 divisions) over two years.

With the parametric model and the exploratory modeling environment, the analyst can now overlay the nominal expansion system on this worry curve. Doing so permits the analyst to read off the number of ready brigades required by the worry curve. Figure S.1 indicates that (in this case, with an expansion system that has three NTCs) the Army would have to maintain not just 30 ready brigades, but a bit over 40 in order to meet *all* the worries on the worry curve. Further, by varying the parameters of the model according to the analyst's ideas of future developments (e.g., advanced training time will be 40 percent longer because of more sophisticated equipment), the analyst can see how an envisioned future would affect the requirements. This interactive capability is a powerful tool for honing one's intuition about expandability.

ACKNOWLEDGMENTS

We are grateful to many people in government, in industry, and within RAND for their contributions to this project. The authors wish to thank all those who contributed their views and expertise. Many people provided us with data, phone interviews, and feedback. Our first thanks go to then-LTC Tim Daniel of DAMO-SSP for suggesting and championing this work. At the risk of overlooking some, we would also like to thank the following: Lou Lypeckyj, Don Livingston, Major Fred Roitz, and Steve Linke of the Abrams Program Office; LTC Rick Ryles of the Apache Program Manager's Office; Victor Burgos and LTC Larry Thomas of the Apache Project Manager's Office; Dave Hegarty for his expert advice on the Patriot; Dan Beck, Patriot Project Office; Dennis Vaugh, MLRS Program Office; Colonel Steve Kratter and Major Max Carroll for their advice on the MLRS; Dave Parabek for his extensive help on Bradley production; Jim Payne, UDLP, FMC Corporation; John Beudy, Bradley Program Office; Dennis E. Mazurek, for data on medium tactical vehicles; Dawn Vehmeier, DUSD (Industrial Affairs and Installations); Connie Tucker and Major Steve Elliott at TACOM; Colonel Wayne Taylor, for his comments on the Army's truck fleet; Tony Currie, George Michaels, and Jean Duval, Cost and Economic Analysis Center, for providing a copy of the FORCES model; Elizabeth Walker and Dwight Schalles at Fort Irwin, for their patience and help with NTC costs; Wendy Freeman at TRADOC, for her data on NTC costs; Lora Muchmore at DUSD (Industrial Affairs and Installations), for fixed cost data; Captain John Auffert, 11th Armored Cavalry Regiment; Terry McLemore, FORSCOM; Quick Kirby, Bob Jordan (MOUT ACTD), Art Durante, and Mr. Christiansen, U.S. Army Infantry Center, Fort Benning; Colonel Scott Armbrister and Captain Mary Martindale, Officer

Candidate School, Fort Benning; Sergeant Wittington, Armor NCO Academy; Major David Collins, Armor School; Colonel Pickens, Major Browder, and Captain Moyer for their hospitality and expert opinions during our tour of the JRTC at Fort Polk; Colonel Patrick Toffler, USMA; Ed Hackworth, TRADOC; Emerson Blake at the U.S. Army Force Management Support Agency, Fort Lee, for providing the 1st Armor division TOE; Mike Dove and Rocky Freudenberg at the Defense Manpower Data Center, for their data on officer accession; Don Durant, Field Artillery School; Colonel Douglas Slater, 2nd Squadron, 16th Cavalry; LTC Russ Hrdy, SARDA, for the tank modernization study; Major Pete Johnson; and Stephen Croall.

Within RAND, the study was improved through conversations with Tom Lippiatt, Tom McNaugher, Jim Quinlivan, John Birkler, Jim Bigelow, Jack Graser, and Gary Massey. In particular, we would like to thank Dick Hillestad and David Hutchison for their patient and thorough reviews.

Finally—and far from least—the presentation of this report owes a great debt to the careful efforts of Laurie Rennie.

Needless to say, any errors or omissions are the responsibility of the authors alone.

ABBREVIATIONS

AC	Active Component
ACR	Armored Cavalry Regiment
AD	Air Defense
AFB	Air Force Base
AIM XXI	Abrams Integrated Management Program for the 21st Century
AMG	Antenna Mast Group
ANCOC	Advanced Noncommissioned Officer Course
AOBC	Armor Officer Basic Course
ARNG	Army National Guard
ARPRINT	Army Program for Individual Training
BIOCC	Branch Immaterial Officer Candidate Course
BLUEFOR	Blue Forces
BNCOC	Basic Noncommissioned Officer Course
Bns	Battalions
BOCC	Branch Officer Candidate Course
BRAC	Base Realignment and Closure
BSNCOC	Battle Staff Noncommissioned Officer Course
CA	Civil Affairs
CALFEX	Combined Arms Live-Fire Exercise

CEAC	Cost and Economic Analysis Center
COL	Colonel
CONUS	Continental United States
CPX	Command Post Exercise
CRS	Congressional Research Service
CS	Combat Support
CSMC	Command Sergeant Major Course
CSS	Combat Service Support
CTC	Combat Training Center
C4I	Command, Control, Communications, Computers, and Intelligence
DI	Drill Instructor
DIVARTY	Division Artillery
DoD	Department of Defense
DRM	Directorate for Resource Management
DUSD	Deputy Under Secretary of Defense
ECS	Engagement Control Station
EIA	Excellence in Armor
EPP	Equipment Power Plant
ERB	Enhanced Ready Brigade
FCS	Future Combat System
FORSCOM	Forces Command
FORCES	Force and Organization Cost Estimating System
FSC	First Sergeant's Course
FTXs	Field Training Exercises
FU	Fire Unit
GDP	Gross Domestic Product
GLDS	General Dynamic Land Systems

HDE	Heavy Division Equivalent
HEMTT	Heavy Expanded Mobility Tactical Truck
HET	Heavy Equipment Transporter
HMMWV	High Mobility Multi-Purpose Wheeled Vehicle
IFV	Infantry Fighting Vehicle
IOBC	Infantry Officer Basic Course
IRR	Individual Ready Reserve
ITB	Infantry Training Brigade
JMNA	Joint Military Net Assessment
JRTC	Joint Readiness Training Facility
LMVS	Lockheed Martin Vought Systems
LTC	Lieutenant Colonel
MANPRINT	Manpower and Personnel Integration
MDHS	McDonnell Douglas Helicopter Systems
METL	Mission Essential Tasks List
MG	Major General
MILES	Multiple Integrated Laser Engagement Systems
MLRS	Multiple Launch Rocket System
MOOTW	Military Operations Other Than War
MOS	Military Occupational Specialty
MOUT	Military Operations on Urbanized Terrain
MP	Military Police
MSA	Minimum Survivable Army
MTV	Medium Tactical Vehicle
MTW	Major Theater War
NCO	Noncommissioned Officer
NG	National Guard
NTC	National Training Center

O/C	Observer/Controller
OCS	Officer Candidate School
OJT	On-The-Job Training
OPFOR	Opposing Force
OPSGROUP	Operations Group
OSUT	One Station Unit Training
POM	Program Objective Memorandum
PLDC	Primary Leadership Development Course
PREPO	Prepositioned
PRV	Plant Replacement Value
PSYOPS	Psychological Operations
QDR	Quadrennial Defense Review
ROTC	Reserve Officer Training Corps
SASO	Sustainment and Support Operations
SMC	Sergeant Major Course
SOF	Special Operations Forces
SRC	Standard Requirements Code
SSO	Small-Scale Operations
TDA	Table of Distribution and Allowances
TOE	Table of Organization and Equipment
TRADOC	Training and Doctrine Command
USAR	United States Army Reserve
USATC	U.S. Army Training Centers
USMA	United States Military Academy
WWII	World War Two
xM	Exploratory Modeling

Chapter One

INTRODUCTION

WHY STUDY THE EXPANDABILITY OF THE 21ST CENTURY ARMY?

From the earliest days of the Minutemen, the United States has been prepared to expand its military capabilities to meet a crisis. In extreme cases, the nation has called upon its industrial might to build additional materiel and has drafted and trained young men and women in the use of that equipment to meet its security needs. The ability to expand our forces was a serious issue during the Cold War. On the heels of the breakup of the Soviet Union and an overwhelming victory in Operation Desert Storm, it is generally conceded that there is no army in the world capable of matching a mobilized U.S. Army on the field of battle. Further, there are few plausible threats to that mobilization capability in the next decade or more. Expandability in the future, then, would seem to be well handled by worrying only about modernizing the Army's forces and ensuring an effective, efficient mobilization plan for current forces held in reserve. Some would argue that the Army's current commitments recommend reconstituting a larger active Army today despite budgetary constraints, but, in general, the Army has more pressing problems. It needs to worry about reducing its mobility footprint, operating better in a joint and coalition arena, dealing with new kinds of threats, modernizing for the information battlefield, etc. The need for studying expandability of the Army is arguably small.

While granting the greater concerns, there are at least three reasons for mistrusting this picture of expandability. The first is that we know so little about today's future. The next 15–20 years hold the potential for significant changes in important national security dimensions. For example, while the national leadership is unconcerned that the geopolitical world will yield a peer competitor in that period, we have only a short experience with the geopolitics of a unipolar world. That is, even in the absence of a peer competitor, our role in the world is still evolving, so it remains difficult to say whether or why we might have to expand or by how much. In addition, there are changes possible that could require a serious rethinking of expandability. The National Defense Panel has argued that "technology, commercial developments, required manpower skills, transnational interrelationships, and the phenomenal expansion of information capabilities bring into question the applicability of traditional mobilization structures."[1] It isn't difficult to imagine, for example, that a "mobilization" of unkempt equipment and "weekend warriors" to an advanced cyberwar doctrine might be much more like a reconstitution in today's thinking than would a "reconstitution" that involved accessing civilian police and their equipment for the expansion of an MP-heavy peacekeeping operation. If there is truly a so-called revolution in military affairs coming, it is not hard to imagine it requiring an expansion capability very different from that of today.

A second reason for not trusting that we know everything we need to know about expandability into the 21st century is the uncertainty in the size of the Army. The smaller the Army gets, the more it needs to concern itself with mechanisms for getting larger, both because it will be easier for a competitor to become a peer and because it will be easier for a security situation to escalate beyond the capabilities of the ready Army. Today the Army is making the implicit assumption that its size has stabilized. But the Army cannot ultimately control its funding or its size, and there is no guarantee that it will not get even smaller in the coming 10–15 years. Sound planning requires the Army to consider the possibility that it could get smaller and to reckon what that would do to—and what the Army should do about—its future expandability. There are already those arguing, for example, that the active Army could get significantly smaller and that

---

[1] *Transforming Defense: National Security in the 21st Century*, Report of the National Defense Panel, December 1997.

the National Guard could be counted on to provide any expansion capability required.[2]

Two separate threads contribute to the third reason for looking at expandability. The first is that "history teaches us a peer competitor will reemerge." This is a phrase heard often from those who are worried that the Army is already too small. History is indeed replete with "superpowers" that were overtaken by emergent "peer" competitors. And typically the superpowers were caught unprepared. Even if few believe a peer competitor *will* emerge in the 20-year planning horizon, the fact that one *could* and the consequences of its happening are sufficient reasons for considering it in strategic planning. Participants in the recently completed Quadrennial Defense Review (QDR) were required to consider "the emergence of a major potential adversary having military capabilities similar to those of the United States."[3] Although both the QDR and the 1998 Posture Statement take the stand that a peer competitor will not appear in the coming 15–20 years, both leave open the possibility.

Even if a peer competitor does not emerge, expansions are common in the Army's history. As shown in Figure 1.1, there have been at least eight significant expansions in the last 190 years, with the longest gap between expansions being about 35 years.

The second thread is that historically, when the United States has expanded forces, it has done so quickly and late. Goldich points to the scrambling expansion efforts required for the 1917 force that went into World War I, the 1941 force that entered World War II, the 1950 force that entered the Korean theater, and even the 1965 force that went to Vietnam.[4] Some point to the Army of Operation Desert Storm as a reversal of this trend, but even that situation was, as Goldich puts it, "a virtual textbook example of the problems of threat identification and appraisal."[5] Even Desert Storm, then, is a stark

---

[2]See, for example, MG (Ret.) F. S. Greenlief, "A Different Vision of the Total Force," *National Guard Review*, Winter 1997, http://www.northupcom.com/winter.html.

[3]From the *National Defense Authorization Act of 1996*, Public Law 104-201, Sec. 924.d.2.E.

[4]R. L. Goldich, *Defense Reconstitution: Strategic Context and Implementation*, CRS Report for Congress, November 20, 1992, pp. 23–24.

[5]Ibid., p. 28.

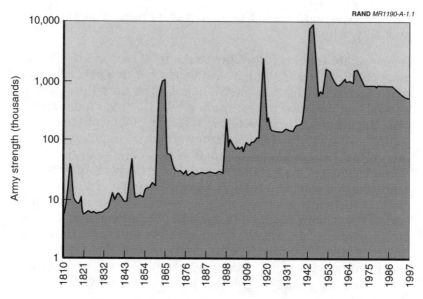

Figure 1.1—Army Troop Strength Since 1810

reminder of the difficulty a representative democracy has in expanding its military capabilities in a timely fashion.

Pulling these two threads together, the possibility (even if small) of an emergent peer competitor and the great difficulty in preparing for a timely and efficient expansion of military capabilities suggest that, if nothing else, Army leaders and planners should work today on the problem of clear and timely indicators and warnings of the need to expand military capabilities.

Finally, in addition to the potential need to take action today to generate or preserve important expandability options, recent advances in computational and algorithmic capabilities (exploratory modeling) provide a new ability to explore the potential impact of future events.

It is based on this reasoning that we have undertaken to study the expandability of the Army in the coming 15–20 years.

STUDY OBJECTIVES

Our primary objective in this study is to help Army leaders and planners think about expandability in a way that most effectively hedges the Army against the possibilities of having to expand in the future. To support that objective, we

- Develop a framework for studying expandability issues today and into the future, and

- Use that framework to explore whether plausible futures suggest action today to facilitate expansion capabilities.

Through exploratory analysis we assess possible future effects on today's expansion capabilities. This provides a better overall understanding of the effects of possible futures on expandability and thus a better foundation from which to make expandability decisions today and in the near future.

Addressing the Issues

The primary issues surrounding expansion of the military forces depend strongly on the planning circumstances. When the call comes for expansion of military forces, the primary concern is how long it will take before the additional forces are mission-ready. In the event it takes longer than planned, as it did with three Army National Guard combat brigades that were called to active duty during the preparations from the Persian Gulf War, post-action issues will include how that preparation time could have been shortened and how much it would cost to maintain that quicker expansion capability.

During peacetime with stable budgets, the issues are broader. They concern the kinds of missions for which the forces should be prepared to expand, how fast they should be prepared to expand, how much it costs to maintain those expansion capabilities, what force structure mix best supports those expansion capabilities, which components should be involved in which expansions (and in what order), with what coalition forces we are likely to expand, and so forth. Most recent studies of expansion have reflected this peacetime and stable (but smaller) budget circumstance.

When budgets are unstable (generally meaning likely to drop) or the focus is well into the future, further issues come into play. Primary among them is the size and makeup of the Army from which the expansion will be done. The further into the future, the more issues come into play about the nature of national security threats and the effect of technological advances on force capabilities.

In these times of unstable budgets and for this study, whose objective is to look 20 years into the future, it would be nice to examine all the pertinent issues. With the exception of issues relating to coalitions, most issues of current concern can be addressed either directly or indirectly. Chapter Two describes in detail which issues can be addressed and how they are covered. In comparison with other studies of expandability issues, then, this study covers more issues and a greater time span, but at a broader level of detail.

On several occasions in its history, the U.S. military has undertaken expansions. The details of those expansions have varied over time, as have the names given to preparations for expansion. For that reason it is important to be clear about what we mean by expandability in this report.

DEFINING EXPANDABILITY

The word most often used in describing an expansion of the U.S. Army is mobilization—a word first used in the 1850s to describe the preparation of the Prussian army for deployment. It was used in the United States during World War II to refer to "the reallocation of a nation's resources for the assembly, preparation, and equipping of forces for war."[6] Today, according to one analyst, "Most military planners argue that mobilization consists of calling up Reserve component units and individuals."[7] In the DoD Master Mobilization

---

[6]Frank N. Schubert, *Mobilization*, The U.S. Army in World War II, The 50th Anniversary, CMH Pub 72-32 (no date), p. 3.

[7]John R. Brinkerhoff, "Reconstitution: A Critical Pillar of the New National Security Strategy," *Strategic Review*, Vol. 29, No. 4, Fall 1991, p. 10. He explains in a footnote (4) that "JCS Pub 1, *Department of Defense Dictionary of Military and Associated Terms* (June 1, 1987), defines mobilization generally but characterizes mobilization categories primarily in terms of Reserve call-up authority."

Plan,[8] the Graduated Mobilization Response defines five levels of mobilization:

- Presidential selected reserve call-up that authorizes the President to involuntarily call up selected reserves for up to 360 days without declaring a national emergency.

- Partial mobilization of ready reserves for up to two years upon declaration of a national emergency.

- Full mobilization of all reserve forces only after Congress has declared a state of national emergency.

- Total mobilization that brings the industrial mobilization base to full capacity to provide additional resources, equipment, and production facilities aimed toward additional units beyond those of full mobilization.

- Selective mobilization of some reserves for domestic emergencies or natural disasters.

The definition of Total Mobilization is close to most definitions of the other major type of expansion in common usage—reconstitution.[9] The strict definition of reconstitution in the 1992 Joint Military Net Assessment is somewhat narrower:

> [reconstitution] is intended to deter a global threat even as we reduce forces by maintaining our nation's capability to establish, equip, and train new units in time to meet a new or resurgent global threat.[10]

One definition of reconstitution differentiates it from mobilization thusly:

---

[8]Office of the Assistant Secretary of Defense (Force Management and Personnel), *Master Mobilization Plan*, May 1988.

[9]Partial, full, and selective mobilization have also been grouped into "regeneration." See Brinkerhoff, p. 11.

[10]*1992 Joint Military Net Assessment (JMNA)*, prepared for the Chairman of the Joint Chiefs of Staff by the Directorate for Force Structure, Resources, and Assessment (J-8), the Joint Staff, August 21, 1992.

Reconstitution need not involve the invoking of emergency authorities usually associated with mobilization, although mobilization and reconstitution may take place sequentially after or before one another. Nor does reconstitution automatically include force modernization, although such modernization can take place concurrently with reconstitution.[11]

This brings modernization—the substitution of new, modern equipment for old—into the mix as well. One report even distinguishes among reconstitution, modernization, and "normal buildup": "actions taken within the framework of the regular DoD planning and budgeting process to accelerate the formation of new force structure faster than would otherwise be the case."[12]

Although the words given to expansion of different types are used inconsistently, differentiating among different types of expansions is useful for distinguishing among the different actions required of each. But in the context of peering 20 years in the future at how long it might take the Army to grow in capability, the various recognized divisions of mobilization, reconstitution, and so forth are more complicating than helpful. Each type of expandability has its own characteristics, but it is those characteristics and how they affect expansion timelines that are important to us, not the label we put on that type of expandability today. For that reason, we will treat *expandability* as *any* capability the Army has (or will have) to grow from its fully ready size and capabilities at a given time to a larger size or set of fully ready capabilities. That is, all means by which the Army can expand will be lumped together under the general rubric of expansion and expandability. Where appropriate, the limitations of this simplification will be presented and discussed.

THE CURRENT STATE OF EXPANDABILITY STUDY

With respect to mobilization, the Army today has worked hard to develop a reserve structure and a variety of mobilization plans geared to expanding military capability in support of national security. The frequency of deployments in recent years has also given the

---

[11]Goldich, op. cit., p. 1.

[12]Ibid., p. 11.

Army extensive practice in the art of tailoring deployments for the idiosyncrasies of a given crisis. Because of the difficulties encountered in preparing National Guard troops for Operation Desert Storm, there is ongoing work on improving postmobilization training of reserves.[13]

The one significant recent mention of mobilization has been in the National Defense Panel's report on transforming defense.[14] Panel members urged a review of mobilization policy to insure its balance, timeliness, relevance, and synchronization. In discussing balance they state that "it is more important to have a weapon on hand in adequate quantities than to have the capability available to produce that weapon six months or a year later." Under timeliness, they state, "Should a hostile peer competitor emerge, then we should make appropriate policy decisions at that time, including mobilization preparation within a sufficient lead-time, in order to be ready if hostilities break out." As for relevance, "In these times of rapid technological advancement, neither stored weapons, materials, parts, nor manpower are necessarily relevant to the mobilization needs of future warfare." And for synchronization, "It makes no sense to have manpower assigned to mobilization units if there is no equipment nor to provide equipment for mobilization purposes without the manpower or without sufficient equipment for active components."

The story with respect to reconstitution is somewhat different. In the immediate aftermath of the Cold War, President George Bush was worried about the possibility of a reemergent Russian superpower. In a speech at Aspen, Colorado on August 2, 1990, he named reconstitution as an important element of his post–Cold War defense policy.[15] Reconstitution was written into the 1992 National Security Strategy,

---

[13]See, for example, Thomas F. Lippiatt, James C. Crowley, Patricia K. Dey, and Jerry M. Sollinger, *Postmobilization Training Resource Requirements: Army National Guard Heavy Enhanced Brigades*, Santa Monica, CA: RAND, MR-662-A, 1996. Here "postmobilization" refers to the mobilization of units after the mobilization orders have been given. In some reports, postmobilization refers to the period after mobilization of reserve units is complete and reconstitution begins.

[14]*Transforming Defense: National Security in the 21st Century*, Report of the National Defense Panel, December 1997.

[15]*Public Papers of the Presidents of the United States. George Bush, 1990. Book II—July to December 31, 1990.* Washington, D.C.: U.S. Government Printing Office, 1991, pp. 1089–1094 [referenced in Goldich, op. cit.].

the 1992 National Military Strategy, the 1992 Defense Planning Guidance, and the 1992 Program Objective Memorandum (POM) Preparation Instructions. There were also studies done by both the Congressional Research Service and Congressional Budget Office dealing at least in part with reconstitution.[16]

Since then, the threat of a re-emergent Russia has gradually abated to the point that both the 1997 Quadrennial Defense Review and the 1998 Posture Statement by General Henry Shelton, Chairman of the Joint Chiefs of Staff, use the same language to describe the likelihood of a peer competitor in the coming years:

> The security environment between now and 2015 will also likely be marked by the absence of a global peer competitor able to challenge the United States militarily around the world as the Soviet Union did during the Cold War.[17]

With decreasing worry about the near-term or middle-term need to build new forces, reconstitution has fallen off the national agenda since 1992.

Interest in expandability capabilities today, then, rests entirely on worries about mobilizing forces from reserves and equipment already in being and on modernizing today's forces. Even out to 2015, the issue of reconstituting forces is of little or no concern.

APPROACH

Our approach is best understood in the context of the analytic dichotomies we make along three separate dimensions of the problem:

- Current versus future expansion

---

[16]Goldich, op. cit., and *Structuring U.S. Forces After the Cold War: Costs and Effects of Increased Reliance on the Reserve,* Washington, D.C.: Congressional Budget Office, September, 1992.

[17]William S. Cohen, *Report of the Quadrennial Defense Review,* May 1997, http://www.defenselink.mil/pubs/qdr/, and *Posture Statement by General Henry H. Shelton, Chairman of the Joint Chiefs of Staff before the 105th Congress Senate Armed Services Committee, United States Senate,* http://www.dtic.mil/execsec/adr98/index.html.

- Heavy versus light expansion

- Expansion timing only versus expansion timing and cost

Current Versus Future Expansion

Describing the Army's expansion capabilities in the future is clearly more difficult than describing them today—more so than during the relative stability of the Cold War world. For all the manifest benefits that have accrued from "winning" the Cold War, it has complicated life for military strategic planners. Gone are the certainties of who the enemy is, what his capabilities are, how he fights, and where he is likely to start fighting. In their place are major uncertainties not only about what the military should be prepared to do in the future, but even about how the military should see its role and how big it ought to be in the future.

What can be known starts from today. With today's force structure and expandability characteristics, a reasonably clear picture can be drawn of how long it would take the Army to expand to a wide variety of capabilities. From time histories of those expansions, a clear picture of today's bottlenecks will be apparent. Those time histories can draw from and be compared with current research on both mobilization and reconstitution for the accuracy of their general portrayal of the expansion capabilities of today's Army. We start, then, with a well-grounded understanding of what an expansion would look like today.

For addressing the manifold uncertainties of the world up to 20 years out, a different approach must be taken. Most of the issues above are amenable to parameterization. That is, each can be dealt with quantitatively and a parametric model can be built around those quantities. Using our understanding of expansion today, we build a parametric model specifically to test the sensitivities of those quantities as they change and interact with one another. In keeping with the greater-breadth, lesser-detail approach of this research, the parametric model will not have great predictive power.[18] Rather, it is

---

[18]For a discussion of predictive and nonpredictive uses of models, see J. S. Hodges and J. A. Dewar, *Is It You or Your Model Talking? A Framework for Model Validation*, Santa Monica, CA: RAND, R-4114-AF/A/OSD, 1992.

a bookkeeping device that is useful for finding "interesting" regions of expansion parameters.

The model lacks predictive power primarily because the correct parameter values cannot be known today. The model accurately keeps track of the effect of a particular parameter value on expansion times and rates and thus is useful for identifying important assumptions, limiting constraints, and unusual outcomes as one varies the model's parameters throughout their nominal ranges.

In this way we can test the "first-order logic" of Army expandability. If there are no unusual outcomes from the parametric model, it will indicate that one's intuition about expansion is sound—that the bottlenecks to expansion won't change significantly with changes in the world, and that they will behave as one would expect them to. If there are unusual outcomes from the parametric model, they can be explored in further detail for their implications about expandability in the situations that gave rise to them.

Heavy Versus Light Expansion

It is canonical in doing analysis on force structure to differentiate between heavy and light forces. In the context of expandability, the primary distinguishing characteristic is the extent to which the industrial base is involved. The industrial base will be involved in any expansion, mobilization, or reconstitution. There is, however, a significant difference in how fast (or at what cost) an expansion can take place if there is a requirement to produce additional units of the "Big Five" equipment items (Abrams Main Battle Tank, Bradley IFV, Apache helicopter, MLRS launcher, and Patriot fire unit). Whether or not these are to be the equipment items of the Army in 15–20 years, the requirement to build substantial new equipment for the forces is what will distinguish "heavy" from "light" forces in this analysis.

The primary value in distinguishing between heavy and light is that the heavy force analysis—because it is the most stressing in terms of both expansion time and cost—will provide a worst-case expandability picture; the light force analysis will provide a feeling for the best-case expansion timelines and/or costs.

Expansion Timing Versus Expansion Timing and Cost

In a crisis, the primary concern of any expansion of military capabilities is likely to be the time required to accomplish it, particularly given the U.S. history of waiting too long to begin an expansion. It is generally during peacetime—particularly in the aftermath of an expansion—that the cost of maintaining an expansion capability is weighed against the speed with which it can be accomplished. In this research there will be distinct objectives for each of these cases.

Expansion timing analysis. As will be described below, the parametric model we used is not a predictive tool in the sense that it was intended to accurately predict expansion times. It is basically a bookkeeping model that keeps track, as a function of the capacity and duration of each stage of the expansion process, of how many units can be readied in a given amount of time. The capacities and duration of each stage are reasonably well known for today's expansion capability. By adding in today's ability to enlarge the capacity of the expansion system, we can reasonably assess the rates of the various enlarged capabilities. As above, this knowledge can be used to develop the parametric model for looking at the future. But what can be said about the future?

Apart from its ability to answer questions related to specific capacities and durations, what can a parametric model say about expandability in the future? As above, there is a "first-order logic" about expandability timing today: the primary bottleneck in expanding the Army today is training troops up to the materiel the Army has on hand; after that (in the heavy force case) the bottleneck is the ability of the industrial base to produce new materiel. The parametric model can be used to try a wide variety of different capacities, durations, and initial conditions in the expansion system to see under what conditions the first-order logic breaks down or changes in important ways. Understanding where one's intuition about expandability could break down in the future is an important output of this part of the approach.

The other intended output of this part of the analysis was to get a better feel for the important regions of outcome space that should be studied in the part of the analysis that involves both timing and cost. For that reason, we selected a wide spread of values for exploring the possible values for the parameters of the expansion system. For example, we allowed the expansion goals to be as high as 40 heavy or light divisions.[19] It is seriously unlikely that we will be called upon to expand the Army that much in the coming 15–20 years. For exactly that reason, however, 40 was chosen[20] to make sure that whatever the Army's expansion goals might be, they are sure to be included in our analysis. This was done as well for the other parameters of the model.

In the study of expansion timing alone, then, we will use the parametric model to do two things:

• Test what could change about our intuition about expandability today, and what could cause it.

• Look across a wide expanse of future expansion possibilities in hopes of narrowing the timing and cost analysis to a smaller subset of the possible outcomes.

Expansion timing and cost analysis. Again, the parametric model is unlikely to predict accurately the costs of future forces and expansion capabilities. What it can do is deal with the relative costs of various force structures and expandability capabilities. Just as there is a first-order logic for the timing of expandability, there is a first-order logic for the costs of expansion: if new equipment is required in the expansion, those costs will dominate the training costs; if not, the costs of expansion are dominated by the recurring costs of the ready brigades. As above, the parametric model can be used to try a wide variety of different capacities, durations, initial conditions, and costs in the expansion system to see under what conditions the first-order costing logic breaks down or changes in important ways. As will be described in the chapter on costs, this leads to a very different and narrow reckoning of costs than is usual.

---

[19]These are actually "division equivalents," and they will be defined further in Chapter Three.

[20]Although it may seem absurdly high, 40 heavy divisions is arguably what could be needed to handle a hot war with an emergent superpower and 40 light divisions is what it would take to bring stability to a peacekeeping situation in a nation of 40 million people.

The model can also be used to look for regions in its outcome space where strikingly different structures and costs make sense. For example, the Army's ability to produce a certain force structure and size in a very short period is likely to require that structure and size to be composed entirely of active forces. On the other hand, if the timeline to expand is very relaxed, various reserve structures will be a more cost-effective approach. Looking for "natural" boundaries between regions of such different structures/costs (and how they might change in the future) will be the goal of the expansion timing and cost analysis.

REPORT OUTLINE

Chapter Two describes the general framework used in this research. Details of the framework for specific pieces of the research are discussed in the section that specifically deals with them.

Chapters Three, Four, and Five deal with heavy force expansion issues. Chapter Three describes in some detail what would be involved in an expansion of heavy forces today. Chapter Four then uses the parametric model abstracted from the details of Chapter Three and explores the timing of expansions of heavy forces in the future. Chapter Five explores both timing and costs (both recurring and expansion) of expanding the heavy forces.

Chapters Six, Seven, and Eight deal with light force expansion issues. Chapter Six describes in some detail the issues involved in expanding light forces today. Chapter Seven uses the parametric model abstracted from the details of Chapter Six and explores the timing of expansions of light forces in the future. Chapter Eight explores both the timing and costs (both recurring and expansion) of expanding the light forces.

Chapter Nine draws general conclusions from the research and suggests actions related to expandability.

Chapter Two

A FRAMEWORK FOR THINKING ABOUT EXPANDABILITY

FRAMEWORK DESIDERATA

The basic objective of this research was to develop a framework in which to think about expanding the Army both today and into an uncertain future. The requirement to think 15–20 years into a very uncertain future calls for a great deal of predictive humility. As described earlier, our approach was to test the future parametrically for the major sensitivities in the time and money it takes to produce a given amount of military capability. This suggests four important characteristics the framework should have:

- Simple, easy to understand

- Robust in coverage of issues

- As quantitative as possible

- Conducive to producing pictures

The turbulent times strain our ability to produce detailed descriptions of the world 15–20 years out. Because of that, a simple, easy-to-understand framework is the best means of engaging in a structured dialogue about expandability issues. It won't be possible to cover all the issues surrounding the Army's ability to expand, but covering the most basic issues will be mandatory. The more quantitative the model (without undue complication) and the more conducive to producing pictures, the more the framework will be useful as a medium for discussing expandability issues over time.

THE BASICS OF EXPANDABILITY

With the four desiderata in mind, the general goal of the Army can be described as making sure that, for all t and Δt,

$$\text{ready}(t,\theta) + \Delta t * \text{latent}(t,\theta) > \text{threat}(t + \Delta t,\phi), \qquad (1)$$

where ready(t,θ) are the military capabilities ready to be deployed at time t (under conditions θ set by other constraints such as political, economic, social, etc.), latent(t,θ) are the reserve or latent military capabilities that can be made ready within time Δt (given conditions θ), and threat(t,ϕ) are the military capabilities of enemy forces at time t (given the circumstances under which the threat presents itself). In words, for any time t (given the conditions θ and ϕ), the Army's ready capabilities should be sufficient at time t and augmentable over any time Δt so that total capabilities will always exceed those of any threat. It is primarily the time factors that will be of interest to this study. This simplification strips away crucial conditions θ and ϕ and elides over matters of how much greater the Army capabilities should be, whether or not they need to be greater continuously or eventually, and so forth. All of this is necessary to make tractable the exploration of expandability options.

A second equation is necessary to make this simple equation interesting:

$$\text{cost}(\text{ready}(t,\theta)) \gg \text{cost}(\text{latent}(t,\theta)), \qquad (2)$$

where cost(x) is the cost of x. That is, the cost of maintaining a ready capability at time t is much greater than the cost of maintaining an equivalent latent capability that could be brought into a ready capability over time. If costs were similar, the best way to maintain equation (1) would be simply to maintain a large ready capability (ready(t,θ)). Since, in reality, doing so is costly, the challenge is to "optimize" equation (1) in such a way that it can be made to hold at minimal cost.

Both the Army's ready capabilities and its latent capabilities have changed over time; have changed dramatically in recent years; and will continue to change in predictable and unpredictable ways in the future. In talking about expandability, it is important to differentiate between the ready or "expand from what?" forces and the latent or "expand with what?" forces. Equations (1) and (2) are clearly gross simplifications of the Army's responsibilities, but they permit that distinction between the forces (both ready and latent) as a function of time t and the potential expansion of those forces starting at a future time and extending over a time period, Δt. The time Δt can be thought of as measuring time after the "flag goes up" or after it has been decided that the Army must be expanded.

The other important aspect of expandability is expanding "toward what?" In equation (1) this is represented by threat(t): the Army must be prepared to overmatch any threat capabilities (including expandabilities). How the framework will be used in this and other regards is discussed below, but before doing so it is important to address what we mean by "military capabilities."

Defining Capabilities

Equation (1) is defined in terms of military capabilities. Few notions are more complex or have caused more heated debates recently than the notion of capabilities. The crux of this research relies on the ability to range parametrically over a large number of cases. This requires a simple model. To keep the model simple, we must keep the definition of capabilities simple. This excludes many important subtleties, but—as will be seen—the effects of many of those subtleties can then be discussed in qualitative terms with the resulting simple model.

One measure of capability is divisions or heavy-division-equivalents. This measure has widely discussed limitations and flaws, but it is the best-known simple measure of the Army's ability to take on a variety of tasks. We will adopt a similar approach (see Chapter Three) with one common modification and one uncommon one. In the former case, we will distinguish between light and heavy forces. In general, it takes much less time to ready light forces than heavy ones, and each type has its strengths and weaknesses depending on its intended mission. Those missions that require mainly light forces (such as many operations other than war) can be discussed using graphs describing the generation of light forces. Those that require significant contributions from heavy forces (more characteristic of major theater wars or a peer competitor) are best gauged using graphs describing the generation of heavy forces. Those that require a mix can be addressed either through the generation of heavy divisions (if, as usual, they dominate the generation time required), or some combination of the two.

The second modification is to report the readiness not of divisions, but of brigades. Advanced training (such as at the National Training Center) is set up around brigades. There is additional training of brigades at the division level before they are mission-ready. For our purposes we will add that division-level training into the advanced training time, but we will report readiness in terms of brigades because that is literally the product of the advanced training sites.

Note, also, that the definition of capabilities excludes the process of shipping the capability to where it is needed and the process of sustaining that capability in combat or in waiting. This simplification ignores the often staggering logistical problems associated with force projection. On the other hand, it deals with the distance-independent problems of force generation. The problems of force projection are left for later study.

AN EXPLORATORY FRAMEWORK

It is common these days to build an analytic framework for a given problem around a computer model. That is the approach we will take, but with an important modification: the model will be embedded in an exploratory modeling environment developed by one of the authors (Bankes). There are several ways to approach analytically the problem of expandability. Most revolve around an optimization of the expansion problem (either analytically or computationally) given a set of constraints. While the pipeline nature of expansion is well suited to these methods, we were after a different result. Rather than optimizing, we were interested in looking across the outcome space for counterintuitive results or "regions." This more exploratory quest was suggested in part by the ability of the exploratory modeling environment to accommodate such an approach. It provides a different way to frame and explore problems such as expandability. This was the primary reason for choosing exploratory modeling for our approach.

The basic framework for thinking about expandability consists of two elements:

- Parametric computer models
- Exploratory modeling environment

Parametric Computer Models

As discussed earlier, the models (one each for heavy and light forces) we will use are derived from a study of the expansion from today's forces. Those are the subjects of Chapters Three and Six, and the simple models are detailed there. For now it is sufficient to describe the models as parametric representations of the two expansion processes. Each variable in the model is given a nominal value derived from the details of expansion from today. In addition, it is given a range with high and low values chosen to represent those values that the parameter could plausibly take on in the coming 20 years. For example, we have nominally 15 enhanced brigades in the force today. In the future we would never have fewer than 0 such brigades, and it is quite unlikely that we would ever have more than 60 such brigades. Although few would consider either of these extremes likely, it makes the point that if all values in between are considered, then all likely futures of that dimension have been considered. The exact parameter ranges and nominal values used in this research are given in Appendix B.

Such parametric models have been used for decades. The primary problem with such models is that if each parameter is allowed to take on n values and there are k parameters, the number of cases to consider is n^k. This gets large very quickly and makes the full exploration of even medium-sized parametric models intractable.

Exploratory modeling takes advantage of modern computing power to address this problem of dimensionality in parametric models in a new way.

The Exploratory Modeling Environment

Exploratory modeling (xM) is a research methodology that uses computational experiments to analyze complex and uncertain systems.[1] Appendix A provides a brief summary of the methodology and points to some recent applications.

For our purposes here, exploratory modeling is also a set of computational tools that permit one to implement the xM methodology. Exploratory modeling takes advantage of computational power to be able to produce quickly any point in the outcome space. This provides a powerful means for exploring the n^k possible states of a parametric model with k parameters, each having n possible values. In particular, it allows for at least two different approaches to questioning the outcome space[2] that we took advantage of in this research:

- Asking traditional questions in a way that permits visualization of the answer in a more immediate, interactive way than is typical.

- Asking different kinds of "what-if?" questions (xM has been described as an automated, interactive "what-if?" capability).

Asking traditional questions. The most important question we want to answer in this research concerns the first-order logic of expandability. To do so, we need to be able to say under what conditions the current bottlenecks to expanding the Army could change. Said another way, we will want to discuss what sets of parameter values lead to different bottlenecks.

This is akin to a traditional sensitivity analysis, but it can be done with xM in a more interactive, visual manner. The outcome of a given set of parameter values can be presented on a computer screen as a two-dimensional plot along with a set of slider bars representing the current values of the parameters that produced that particular graph. The power of xM is that those slider bars can then be moved to different values of the parameters, changing the graph in response. This provides a very powerful capability to discern what

---

[1]See Steven Bankes, "Exploratory Modeling for Policy Analysis," *Operations Research,* Vol. 41, No. 3, 1993, pp. 435–449.

[2]This doesn't do justice to the power of xM, but it represents those capabilities that are most supported by computational tools at this point and that are most useful for the kinds of questions we sought to answer about expanding the Army.

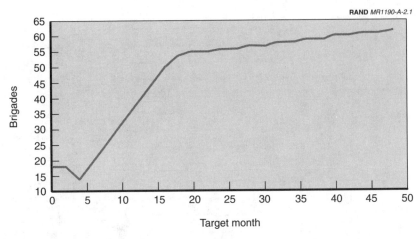

RAND *MR1190-A-2.1*

Figure 2.1—Sample Graph from The Exploratory Modeling Environment

values of the parameters most contribute to changes in the outcomes of the model and what changes occur. A sample graph and sample set of slider bars are shown in Figures 2.1 and 2.2, respectively.

This capability of the xM tools will be used to explore the first-order timing and costing logic of expandability.

Asking different kinds of questions. The ability to access quickly any point in the outcome space allows the analyst to ask some very different questions of the model outcomes. We took advantage of this in the work on costs. In particular, we were interested in looking at points in the two-dimensional space defined by the target size of an expansion and the time required to achieve that size. For each point in that space—presuming perfect knowledge of both how large a force is required and how much time is available for accomplishing the expansion—an "optimum" path can be computed (optimizing, for example, by minimum recurring costs). Each point in space is then defined by a force structure, and the space can then be literally "colored" in a variety of ways (for example, by how many enhanced ready brigades are in that optimum force structure). Then one can look at two- and three-dimensional subspaces for interesting regions

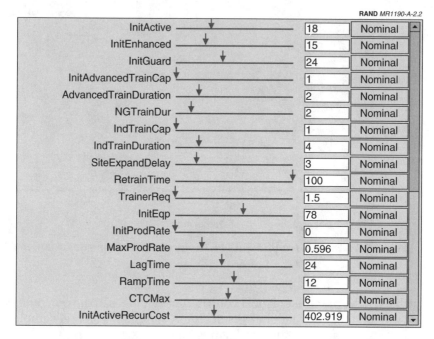

RAND *MR1190-A-2.2*

Figure 2.2—Sample Set of Slider Bars

and boundaries.[3] Changing slider bars related to the components of, say, recurring costs, allows the analyst to watch how those changes affect the colored regions and boundaries. Such a capability is particularly useful for our goal of describing the future of costing bottlenecks: it is how the colored regions in size-time space move as parameter values change that is of interest, rather than the details of an optimal expansion path.

Figure 2.3 is an example of a colored region plot. In this case, the colored regions represent bands of recurring costs for reaching a target number of readied brigades in a targeted amount of time.

---

[3]See, for example, R. J. Lempert, M. E. Schlesinger, and S. C. Bankes, "When We Don't Know the Costs or the Benefits: Adaptive Strategies for Abating Climate Change," *Climatic Change*, No. 33, 1996.

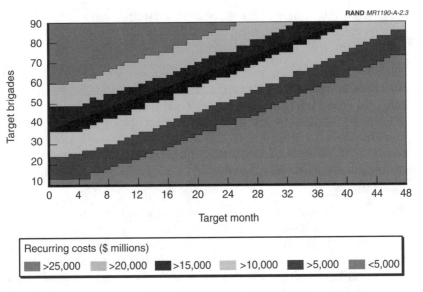

Figure 2.3—Sample Colored Region Plot

Documenting Exploratory Modeling Outcomes

Before proceeding, a brief mention of documentation is in order. Because of the very visual, interactive nature of xM, the documentation of the results is somewhat problematic. It is one thing to notice—while sitting at a computer screen—that moving a slider bar does little to change a given graph. It is another to demonstrate that observation to the satisfaction of a critical reader. That challenge will be taken up in Chapter Four.

ADDRESSING EXPANDABILITY ISSUES

Description/Exploration Versus Prescription

The framework represented by equations (1) and (2) is general enough to apply to the past, present, and future. It is important to state again, however, that its powers are largely descriptive rather than predictive. The distinction between descriptive and predictive goes to the question of validation of the parametric models. They are not *validated* models of the expansion system. They cannot validly predict the advanced training time given a set of expansion facilities, trainers, doctrine, equipment, etc.

They are *verified* models of a bookkeeping system in the sense that they have been verified to keep the books accurately. They can compute, say, the total training time (or cost) if each of the times (costs) for the individual training segments that make up training are known. More important, with the ability to vary those training times (costs), the models can compute and track the changes in total training time (cost) as a function of the changes in the individual segments.

The models cannot be used to predict future training times, costs, or capabilities. They can be used to keep track of how future changes in the times, costs, or capabilities of individual segments will affect the overall system. They can thus be used to reason about how future worlds will affect the training system and to suggest qualitative actions that can improve the expansion system.

Developing Capabilities over Time

The right side of equation (1) deals with the threat. Exactly how the threat enters into the research is described below, but it is fair to say that the nature of the threat will be largely overlooked—not because it is unimportant, but because it is difficult to quantify and will therefore be largely subsumed into the target size of the expanded Army. One can argue that both the ease with which the Army could expand to overmatch any extant threat and the difficulty in predicting future threats make it preferable to concentrate the research on the left side of equation (1). This is also in keeping with the defense community's current concentration on capabilities rather than threats. But there are risks associated with largely ignoring the threat and at least one is worth spending some time discussing.

How fast one can generate forces is tied closely to the specific nature of the threat. The shortest or optimal time required to generate a given number of brigades/divisions depends strongly on the number desired. That is, if two divisions are needed, the optimal time to generate those two divisions will generally be shorter than the time to generate two divisions if one is trying to optimally generate a larger number (say 10) of divisions. If two divisions are needed, they can be readied and shipped. If 10 divisions are needed, it might be possible to do it faster if nearly ready divisions are not readied but instead used to help generate and train the other eight divisions. This means that the first two divisions of the optimally generated 10 divisions will not be ready to ship as quickly as if only two divisions had been needed.

To presume that a large number of units can be generated optimally is to fly in the face of U.S. military history. The Army has typically had to send forces somewhere quickly and generate more as best it could. For the purposes of this analysis, however, we will presume in all cases that the generation of a large number of forces will be done in a roughly optimal way. In practical terms, this means that there will generally be fewer brigades/divisions ready for deployment immediately than ready troop strength would suggest, because the optimal approach would use trained brigades as trainers to increase the training rate.

Expandability Issues

The framework can be used to address some, but not all, of the issues that concern people about expandability. The issues can be broken down roughly into four categories:

- Expanding from (and with) what?

- Expanding toward what?

- How?

- Cost?

It is important to be clear at the outset which issues can and cannot be addressed in each category.

From (and with) what? Any expansion of Army capability is built on the forces, materiel, and industrial base in place at the start of the expansion. As military budgets have been dropping, each of these starting conditions has undergone significant change in size as well as composition. The framework will be able to address the expandability effects of a variety of structures and mixes of the ready and latent (reserve) forces by varying the time each requires to prepare for deployment and the mix of each at the start of expansion. There are many subtleties of expansion from the various components that the framework will not do a good job of capturing. The framework will generally assume that the readiest reserves will be trained next without regard to subtleties as to the types of reserves needed for a given expansion. Such issues could be explored indirectly by varying the readiness of specific reserve units, but we did not do that in this research.

Much of the materiel available today will be usable—if aged—out 15–20 years. Changes to that materiel are likely to be in the form of improvements rather than new equipment. With respect to materiel available, the framework will generally follow current equipment tables and improvement schedules. The ability to introduce new equipment will be handled by being able to vary the availability of an industrial base to produce that equipment from warm or cold production lines, a number of production shift capabilities, and a speed at which new equipment can be produced.

Toward what? An actual expansion of Army capabilities will be driven by the characteristics of the specific need. These include the mix of required forces, the nature of the threat, the availability of and requirement for joint and/or coalition operations, the urgency of the situation, and so forth. The sheer volume of potential characteristics and the uncertainty of their appearance precludes a detailed study of the threat or "toward what?" The mix of generated forces will be restricted to two cases: all heavy forces and all light forces. This will bound the timelines for expandability and permit rough comparisons between expanding for major theater wars and lesser contingencies. The ability to generate intermediate mixes of forces can be inferred indirectly from the two bounding cases.

The ability to explore expanding toward an "information age" Army or some other futuristic force requirement is similarly constrained. The demands of such forces, however, can be explored parametrically in the sense that having a variety of times required for producing equipment or times required to train to equipment will illuminate indirectly the effects of equipment that places that sort of demand on expansion capabilities.

The issue of expanding for joint operations is difficult to address directly, in that forces from the other services are not included in the study. The issue can be addressed indirectly in the sense that if one has a notion of how joint operations will affect the size, training time requirements, or other parameter that is modeled, one can address the effect of joint operations by testing it through its effect on the modeled parameter. If, for example, one feels that joint operations will increase the advanced training time required by 50 percent, one can look at the curves in the results where the advanced training time has been increased by 50 percent.

The issue of expanding in concert with a coalition is one that must be forgone entirely. Not only are there too many possible coalition arrangements to explore, there aren't good parametric ways of introducing the effects of such preparations. In this report, then, we are assuming unilateral expansion of U.S. capabilities.

How? Primary among the issues of how one expands military capabilities is "how fast?" The capability to expand quickly is tempered by the cost to maintain the forces at levels from which they can be expanded that fast.

There are two major issues with respect to training: expansion and peacetime training requirements. Expansion training requirements can be dealt with directly through the framework by permitting exploration of the varying times to complete predeployment training. Peacetime training requirements will be dealt with only indirectly in that they will be reflected in the readiness of the various components at the time expansion begins.

Exploring the balance between force structure and modernization is more difficult to address. The effects of current improvement plans can and will be presented. Tradeoffs between force structure and modernization can only be addressed indirectly, through the costs of "improvements." Current modernization plans affect the availability of the industrial base for generating new equipment, and that can be addressed directly.

Cost? Maintaining ready and latent capabilities over time requires active expenditures. These involve expenditures for training, equipment maintenance, producing materiel from current production lines, maintaining "warm" production capability, research and development, etc. These data will be used to parameterize the costs of those capabilities in the model, and the exploratory modeling can then assess the effects of cost shifts on the major expansion constraints.

SUMMARY

The primary intent of the framework is to explore the major expansion constraints today and into the future. There are several expandability issues that could affect those expansion constraints. The framework has the following capability to address those issues and their ability to affect the major expandability constraints.

Issues the framework can address directly:

- Future force structure and mix
- Readiness
- Industrial base capabilities
- Mobilization training requirements
- Expansion timelines
- Cost

Issues the framework can address indirectly:

- Expanding from which component
- Future requirements, threats
- Effects of information age, other future technology
- Balancing force structure and modernization
- Peacetime training requirements

Issues the framework cannot address:

- Expansion with coalition partners

Chapter Three

EXPANDING THE ARMY'S HEAVY DIVISIONS TODAY

Force expansions today would differ depending on the final force level goal and the time available to reach that goal. Without a specific extant threat that might require an expanded Army, we have chosen to look at expanding the force up to 40 heavy divisions (120 brigades) under both current and expanded training and equipment production capabilities. An expansion this large exercises all the elements of the expansion system—and this is the primary justification for looking at it.

Our concern at this point is to explore how long it would take to expand to 40 divisions today with various expansion capabilities and to describe the processes involved. The general expansion process can be characterized by the flow model outlined in Figure 3.1. Although the actual process has been greatly simplified in the model, the major elements involved in training and equipping the force are represented.

This flow model represents three kinds of elements: (1) the individuals or units that are transformed into ready units by the training process; (2) the training process, facilities, and training personnel; and (3) the production factors that are transformed into equipment by the production process.

The first element in the expansion process is the individuals or units that receive the training. There are four types of forces that may receive training: (1) active component forces; (2) enhanced ready brigades (ERBs); (3) Army National Guard (ARNG) brigades; and (4) civilian training brigades. As shown, each enters the expansion process in a different place, depending on its readiness for training.

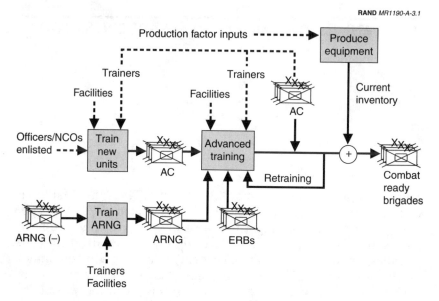

RAND *MR1190-A-3.1*

Figure 3.1—Flow Model

The second element—the training processes—is represented in Figure 3.1 by the three training boxes. To train new units, enlisted personnel receive basic and MOS (military occupational specialty) training, and NCOs (noncommissioned officers) and officers are trained or brought in from active units to lead them. When this training is completed, active component (AC) units are formed that are ready for advanced training. ARNG units would have to fill individual and unit deficiencies before they were ready for more advanced training. Advanced training at the battalion, brigade, and division level is the final training for all units and must be completed before units are combat ready.

The process of training requires training and support personnel, as well as facilities. Six types of training and support personnel are required: trainers, training management personnel, training support personnel, installation and higher-echelon support personnel, and, for advanced training facilities, simulation support personnel and opposing forces (OPFOR). A sufficient number of all these types of training personnel must be available for effective training. In the model, the AC forces supply these trainers if they are otherwise unavailable.[1] In addition, training facilities of different types must be available, depending on the type of training required.

The third element in the expansion process is the production factors transformed by the production process into equipment. This will include both the status of the industrial base today as well as its current ability to expand to greater production.

The first three major sections below take up the three main elements of the expansion process. Each includes both the status of the system today as well as its ability to grow were a large expansion called for. The fourth section describes the maximum ability of the expansion system to provide for trained and equipped troops.

TRAINEES

There are currently not enough active and reserve forces to expand to 40 divisions. This expansion will require not only all of the current reserve forces, but new units as well. The various unit types and the training requirements for each are taken up separately, beginning with the active units.

Active Component or "Ready" Brigades

The AC brigades are nominally ready for deployment. They maintain individual skills as part of their daily routine and currently receive advanced training at the National Training Center at Fort Irwin about every two years. This occasional retraining of active brigades ensures that all active forces remain well trained and at a high state of readiness.

Enhanced Ready Brigades

ERBs are the next type of force that would be readied in an expansion. They are the reserve component with the highest peacetime readiness level. Because of the perceived slowness of some ARNG

---

[1]We assume that IRR (Individual Ready Reserve) and ARNG trainers may be needed as fillers to mobilized ARNG divisions and for ARNG training.

units to get ready for combat during the Persian Gulf War, 15 ARNG brigades have been designated by the Army as ERBs. The intent is to keep these units at a C-1 level of readiness, which means that 90 percent of the personnel are present and qualified and all equipment is available and operational by the time training begins. ERBs should be ready within weeks of the beginning of an expansion to go to advanced training.

ARNG Brigades

ARNG brigades are the next type of force to be readied. In Figure 3.1 these forces are represented by "ARNG(–)," which indicates some deficiencies, and then by "ARNG," which represents the same forces with their deficiencies corrected. After the correction, these forces are ready for advanced training. In addition to the 15 ERBs, the ARNG has eight more heavy divisions (or at least division flags) as of this writing. The C-rating of these eight divisions varies, but it is, in general, considerably below the C-1 ratings of the ERBs.

The typical deficiencies of an ARNG unit are of two types: (1) insufficient numbers, primarily among enlisted personnel, of particular MOSs, and of NCOs and officers in general; and (2) deficiencies in the training of those who are available. The training will be discussed in the next section.

The personnel to fill the various enlisted, NCO, and officer vacancies would probably come from the Individual Ready Reserve (IRR). The IRR constitutes the largest of the Army's pretrained individual manpower categories. IRR personnel provide the primary source of fillers required by the AC, the ERBs, and particularly the ARNG divisions.

There are currently about 202,164 IRR enlisted personnel, 31,518 IRR NCOs, and 56,569 IRR officers.[2] Of course, requirements determine which specialties are needed, but eight ARNG divisions and five ERBs would require only a total of about 260,000 enlisted personnel and NCOs, and about 52,000 officers. The number of available IRR personnel is probably sufficient to fill the ARNG deficiencies unless the

---

[2]Current as of 1997. *Strength of the Army, Part III Strength*, Reserve Components USAR, HQ, Department of the Army, U.S. Army Reserve Personnel Center, St. Louis, May 31, 1997.

number of ARNG units grew considerably larger than today, especially because IRR personnel can be retrained, if necessary, in a new specialty.

New Units

Finally, civilians are the last type of force to be readied. Enlisted civilians first receive basic and MOS training, get a complement of NCOs and officers (who might also need training), and emerge as an active component unit. Newly formed units would require the accession of enlisted personnel, NCOs, and officers.

Enlisted personnel. A heavy division consists of about 20,000 personnel, of which about 85 percent are enlisted. These 17,000 required enlisted personnel would have to undergo basic training and MOS training before they were ready to form units. If the capacity of 6 NTCs (National Training Centers) is about 9 divisions per year or 18 divisions per year with 12 NTCs, then as many as 150,000 to 300,000 enlisted personnel would have to be trained each year to meet these capacity requirements of the advanced training system.

Noncommissioned officers. NCOs comprise the backbone of the Army today; they are the reservoir of experience that enables 21-year-old second lieutenants to effectively command troops. Modeling the rapid expansion and accession of experienced NCOs is difficult. Competent college graduates can be taught the fundamentals of infantry tactics along with the special administrative tasks associated with the platoon leader job. But evaluating exactly how much and what type of experience a typical E-6 staff sergeant requires is difficult.

Some NCO specialties and ranks are more difficult to fill than others, and they are more critical for the smooth operation of the infantry and armor branches. It is useful to examine, at least briefly, the various NCO specialties and the formal NCO courses that are given at various stages throughout an NCO career to determine which of these ranks and specialties would be the most difficult to fill in an expansion.

U.S. Army NCO Academies for Infantry and Armor are currently located at Fort Benning and Fort Knox, respectively. The NCO courses are the Primary Leadership Development Course (PLDC), Basic NCO Course (BNCOC), Advanced NCO Course (ANCOC), Sergeant Major Course (SMC), the Battle Staff NCO Course (BSNCOC), First Sergeants Course (FSC), and the Command Sergeant Major Course (CSMC) (see Table 3.1).

The lowest NCO command is the fire team, typically held by a specialist, corporal or sergeant (E-4 or E-5). The PLDC course trains E-4s to assume the responsibilities of this command level. PLDC courses are ubiquitous and easily set up. During the Vietnam War, raw recruits were molded into instant sergeant E-5s or "shake and bake" NCOs. Promising recruits were identified in basic training and eventually enrolled in the Excellence in Armor (EIA) program, a 90-day intensive course.

The platoon sergeant position is absolutely vital, but NCO commands at the E-7 command level are not of concern because the available pool of both active and reserve lower-ranking E-6 NCOs is large enough to promote from within the ranks. If there are four experienced E-6s to choose from for every newly required E-7 slot, enough experienced personnel should exist to meet this need. The same rationale holds true for NCO ranks above E-7.

Table 3.1

NCO Requirements

NCO Academy Course	Typical Trainee (Pay scale / Rank)	Future Armor Job	Future Infantry Job
PLDC	E-4 or E-5 / corporal, specialist promotable	Team leader	Team leader
BNCOC	E-5 or E-6 / sergeant promotable	Tank commander	Squad leader
ANCOC	E-6 or E-7 / staff sergeant promotable	Platoon sergeant	Platoon sergeant
FSC	E-7 sergeant 1st class promotable	Company 1st sergeant	Company 1st sergeant
SMC	E-8 master sergeant promotable	Primary staff NCO at Bn HQ	Primary staff NCO at Bn HQ

We believe that the greatest problem could come at the E-6 level for the squad leader/tank commander position, which requires skill and several years of experience. There are 58 tank commander positions in a typical tank battalion. Half of these positions are filled by platoon leaders (O-1s) or platoon sergeants (E-7s), half by E-6s. A new heavy division would require about 160 E-6 tank commanders. Therefore, about 960 to 1,920 E-6s would be required each year of an expansion to meet the demand.

Officers. During a national emergency that resulted in a major force expansion, more officers of all grades would be required. However, through promotion and a call-up of the IRR, most senior officer grades could probably be filled. In World War II, the most critical shortage of officers was at the most junior rank—second lieutenants.[3] In another major expansion, O-1s would again probably be most in demand, because many new O-1s would be required and there are no lower ranks from which to promote. Because this chapter focuses on the expansion of heavy divisions, we will focus primarily on the accession of armor and infantry O-1s and not other specialties (such as field artillery and aviation), which are also necessary for forming heavy divisions. Armor and infantry officers comprise the majority of officers in an armored or mechanized division.

In heavy divisions (armored or mechanized), every platoon is led by a second lieutenant, which gives a requirement of about 82 armored second lieutenants per heavy division. In addition, there are about 102 infantry platoon leaders required per division.[4] If 9 to 18 divisions are required to begin advanced training at 11 months, then about 738 to 1,476 armor second lieutenants and 918 to 1,836

---

[3]This was especially true for the infantry branch. For simplicity, we assume all platoon leader positions are held by second lieutenants (O-1s), even though this command position can also be held by first lieutenants (O-2s).

[4]Here we assume there are three platoon leaders in each line company. There are four line companies and one HQ company per mechanized infantry battalion, and 5.67 mechanized infantry battalions per average heavy division (assuming there are 2 mechanized divisions for every armored division in the force structure). In addition to platoon leader slots, there are at least three other various positions for infantry lieutenants in battalion mortar, scout, and anti-armor platoons, or as staff positions at battalion level: (3 Plts/Co * 5 Co/Bn) + (3 Special Plts or staff/Bn) * 5.67 Mech Bn/Heavy Div = 102 infantry officers. The same basic logic was used for the number of armor lieutenants required for a heavy division: (3 Plts/Co * 5 Co/Bn) + (4 Special Plts or staff/Bn) * 4.33 Armor Bn/Heavy Div = 82 armor lieutenants.

infantry second lieutenants would be required, with the same number required every year thereafter.

TRAINING PROCESSES, FACILITIES, AND PERSONNEL

The most demanding of the training processes is advanced training for brigades. We will describe it in some detail before dealing with the other training concerns.

Advanced Training[5]

In a major force expansion today, we assume that all units would have to undergo advanced training at a combat training center (CTC) before they were ready for combat. Currently, in peacetime, individual soldiers join existing units after receiving basic and MOS training as personnel fillers and can enter combat without advanced training.[6] In a major force expansion, however, most new soldiers would not fill out existing units but would form new units. In a national emergency, these new units could be certified as combat ready without advanced training. Even in World War II, soldiers underwent progressive unit and combined arms training after basic and MOS training before they were sent into combat. Since combat operations are much more complex today (and are likely to be even more so in the future), we assume advanced training is required for unit combat effectiveness.

Activities and organization of advanced training. The activities that a unit must complete depend on the missions for which it is being trained. Here we assume that these units will be trained to execute

---

[5]Two recent RAND studies have focused on the advanced training process: Thomas F. Lippiatt, James C. Crowley, Patricia K. Dey, and Jerry M. Sollinger, *Postmobilization Training Resource Requirements: Army National Guard Heavy Enhanced Brigades*, Santa Monica, CA: RAND, MR-662-A, 1996; and Thomas F. Lippiatt, James C. Crowley, and Jerry M. Sollinger, *Time and Resources Required for Postmobilization Training of AC/ARNG Integrated Heavy Divisions*, Santa Monica, CA: RAND, MR-910-A, 1998. To analyze the advanced training resource requirements and to estimate training times, we will borrow from these analyses.

[6]In 1993, 175 brand new infantry privates were assigned to the 10th Mountain Division and sent to Somalia. They ended up in the firefight in Mogadishu on October 3, only three weeks after finishing their OSUT (one station unit training).

three major missions: movement to contact, deliberate attack, and area defense. These missions were chosen because they are essential for major theater war.

If a unit is trained for these missions, then it must complete a series of activities—from individual to division-level training—before it is ready for combat. Table 3.2 identifies training activities for the brigade and the approximate time allocated for each activity.

The simplest estimate of the training time for a brigade to complete these required activities is a total of 102 days. However, the steady-state flow rate through the advanced training process is faster than this sum because a given unit does not occupy the entire training facility all the time it is there. Units can share some portions of the training site with other units, which increases the steady-state flow rate of units through the facility.

The training activities can be organized in a variety of different ways to share the training facilities. Different modes of organization have different virtues, but we have chosen to focus on the fastest option.[7] Our model also accounts for division-level training, which results in a lower flow rate than a simple brigade-level flow rate.

Table 3.2

Advanced Training Activities

Activity	Description	Days
Initial prep	Initial preparation and movement to training site	17
Gunnery-and-below training	Individual, squad, platoon, and gunnery training	33
Company-level training	Task force organization and company training	17
Brigade-level training	Battalion task force and brigade-level training	25
Final prep	Maintenance, equipment services, final preparation	10
Total		102

SOURCE: Lippiatt et al. (1996).

---

[7]Lippiatt et al. (1996).

Advanced training inputs. This advanced training process must have adequate facilities and a sufficient number of capable trainers to effectively conduct advanced unit training. First we describe the required facilities.

- **Requirements.** Advanced training has a number of requirements for the facilities where such training is to occur if the training is to be effective. Possible training sites must have the following:

  — *Gunnery range.* To be considered for advanced training, a site must have sufficient gunnery ranges to conduct gunnery exercises.

  — *Combined arms live-fire exercise (CALFEX) capability.* A site must have sufficient space to conduct two simultaneous company-level exercises.

  — *Maneuver space.* A site must have enough maneuver space for two simultaneous battalion-level force-on-force exercises.

  — *Availability of facilities.* Sites must be available and not used by AC units.

  — *Installation facilities support.* Sites must have adequate facilities to support the training.

  Currently the National Training Center meets these criteria and is the Army's operational advanced training facility.[8] However, there are other potential facilities that could be used in a national emergency.

- **Expansion.** Based on the above criteria, the Army potentially has the following sites available for advanced training: Forts Hood, Bliss, Carson, Irwin, Yakima, and Gowen Field.[9]

  In addition to these facilities, the Army has two other large land tracts: White Sands and Yuma. White Sands has about 2 million acres of land, and Yuma has about 1 million acres. Using Fort

---

[8]It is the only CONUS-based Combat Training Center (CTC) for heavy units.

[9]Gowen Field does not now have sufficient maneuver space but possibly could acquire more.

Irwin, with about 600,000 acres of land, as the basis for determining the potential size required for building training sites, perhaps three training sites could be constructed at White Sands and two more sites at Yuma. Furthermore, Fort Bliss (1.2 million acres) may be large enough to accommodate two sites.

Both White Sands and Yuma would meet the area criteria identified above to conduct various exercises, but neither has installation facilities support. Nevertheless, in a national emergency, installation facilities could be built. Perhaps these facilities would not be as sophisticated as those at Fort Irwin, but in a three-month period, we estimate that adequate facilities could be built, at least to begin training. The Army currently has access, then, to land that would support up to 12 NTCs.

Now we describe the required training personnel for the advanced training process.

- **Requirements.** Several types of training personnel are required to operate an advanced training site. Trainers control the exercises and observe and record the results of the exercises; training management personnel plan and coordinate the exercises; training support personnel conduct many miscellaneous activities, such as generating smoke and preparing fires; simulation support personnel run and maintain the command and control simulations; and installation and higher-echelon support personnel provide administrative and logistical support for the ARNG brigade.

  The opposition force (OPFOR) provides brigade-level force-on-force opposition for each training brigade and is the highlight of the training activity. It is trained to operate as an enemy force using enemy doctrine and tactics. Currently, the 11th Armored Cavalry Regiment (ACR) provides the OPFOR at Fort Irwin.[10]

- **Expansion.** If the Army were to expand the number of its training sites, it would require substantially more training personnel. Lippiatt et al. (1996) identified a personnel requirement of train-

---

[10]Usually, engineer and infantry companies also augment the ACR, as well as the 1/22 Nevada National Guard.

ers, training management, and training support of about 1,000 trained personnel per NTC.

Trainers and training managers would have to be experienced in the particular grade and MOS that they were observing and controlling. For this reason they would have to be supplied from the active component. Training support would need less specific training and could come from the reserves. There are enough potential training personnel in the current reserve component and active TDA (Table of Distribution and Allowances) force to man about three NTCs.[11]

Expanding to more NTCs would require the use of active component TOE (Table of Organization and Equipment) forces. Although it is difficult to estimate the number of active TOE forces required for trainers and training managers, we have assumed for our nominal estimate that perhaps one active heavy brigade could provide the trainers necessary to man two NTCs.

The OPFOR force at Fort Irwin is currently the 11th ACR. But any active heavy brigade could provide an adequate OPFOR with some training in potential adversary tactics. Thus, one active brigade per NTC would be required to provide an OPFOR.

Combining these active force requirements, we estimate that the first three NTCs could be manned with current reserve component forces and current TDA Army forces. Each additional training site would require about 1.5 active heavy brigades to man.

Maximum capacity. Using two training sites, the estimated maximum rate of advanced training is two divisions in 239 days and two more divisions in 200 additional days.[12] If we assume that sufficient training personnel are available and that building new sites requires about three months, then the capacity of 3, 6, and 12 NTCs to train combat ready brigades is shown in Figure 3.2.

For example, 6 NTCs can meet a cumulative training demand of 18 brigades by the fourth month and 36 brigades by the eleventh month.

_____

[11]Lippiatt et al. (1996).

[12]Ibid.

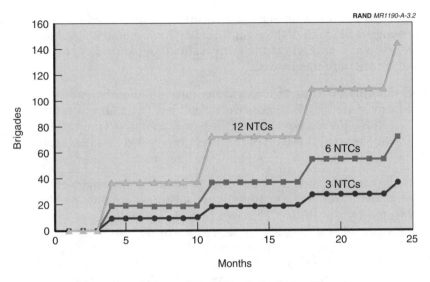

Figure 3.2—Advanced Training Cumulative Capacity

Training Up for ARNG Deficiencies

Most IRR troops needed to fill out personnel deficiencies in an ARNG unit would require additional training to refresh their skills. During a partial or full mobilization, all the U.S. Army Training Centers (USATCs) and schools would go into operation to provide refresher courses not only for IRRs, but also for the present ARNG personnel who were found to be deficient.

IRR refresher courses last about four weeks. The number taking these refresher training courses could be greatly expanded during a full mobilization. Noncritical classes would be eliminated; course size would be increased; nontraining periods would be reduced or eliminated; and the number of training hours per day and the number of training days per week could be increased. It is difficult to estimate the additional capacity of this training system, but it could probably be expanded to meet requirements.[13]

---

[13]*TRADOC Mobilization and Operations Planning and Execution System 1-97*, Annex T, HQ, U.S. Army, Training and Doctrine Command, Fort Monroe, VA, May 30, 1997, p. T-3-3.

These ARNG divisions have a division structure in place, and with enough time, each of the eight ARNG divisions could fill its divisional MOSs with enough sufficiently skilled personnel to begin the advanced training process.

Estimating the time to make these units sufficiently ready to begin advanced training is a controversial subject. The Army's estimate is that ARNG divisions would require 9 to 12 months to prepare for war. For this reason, these divisions are not included in any current war plans.

Using our flow model, rather than estimating the length of time to prepare for war, let us estimate when advanced training sites will be vacated by the ERBs. This will provide an estimate of when the ARNG needs to be ready to occupy these training sites and begin advanced training. The total time to get ready for advanced training and the time for that training will provide an estimate of the practical time to train these units to a wartime readiness level in a pipeline model.

If all ERBs were trained at once as divisions, they would occupy five NTCs. If the ERBs have about 90 days before trainers and NTCs are available, and then require about 239 days to train, they will leave their respective NTCs in about 329 days. However, about 39 days before they have fully completed their training, they will have vacated some of their training facilities so that new units could enter to begin to train.[14] Thus, the ARNG divisions would have about 290 days, nearly 10 months, before they could begin advanced training at these sites.

Given that there are probably enough IRRs to fill the ARNG shortfalls, even by Army estimates 10 months is enough time to remedy their deficiencies and make these units ready for advanced training.

If only five NTCs were operated, the ARNG would be ready to occupy these sites when they were vacant. But if more NTCs were opened, would ARNG units be available early enough to fill additional sites?

---

[14]Lippiatt et al. (1996).

If the Army's estimate of 9 to 12 months for ARNG divisions to be ready to deploy (including an NTC rotation) is correct and the time to train the first divisions at the advanced training sites is about eight months, then the ARNG units would have about one to four months to prepare for this advanced training.[15]

If NTCs require three months to become operational, then probably enough ARNG divisions would be available to fill these sites. Our nominal estimate is two months for ARNG divisions to be ready to begin advanced training.

New Unit Training

Enlisted personnel. All enlisted personnel must receive Initial Entry Training (IET), the introductory training given to all enlisted personnel upon entering the Army. IET consists of Basic Combat Training (BCT), which focuses on soldier skills, and Advanced Individual Training (AIT), which qualifies a soldier in a MOS. For some MOSs, soldiers stay at their USATC, and for others, they would be transshipped to other facilities. One station unit training (OSUT), which combines BCT and MOS qualification training into one course, is conducted at one installation.

Currently, 6 peacetime USATCs and 11 other MOS training facilities are in operation. These USATCs and schools currently operate at a rate that produces an adequate supply of trained enlisted personnel for peacetime operations (about four brigades every four months).[16] However, this rate would not be enough to fill the requirements of a major expansion.

Under a major expansion, the current capacity could be greatly increased. The training tempo could be increased with the greater use of all training facilities. Class sizes could grow, hours of use could increase, and the rate of training at current USATCs and

---

[15]Some estimates are as low as four months to prepare an ARNG division for combat, but apparently these estimates do not include the advanced training.

[16]This graduation rate varies depending on whether it is based on basic and MOS infantry graduation or Armor graduates. In 1997, Fort Benning graduated about 13,000 11Bs and 11M infantrymen. Fort Knox is projected to graduate 3,747 19K armor crewmen and 1,687 19D cavalry scouts in FY98.

schools could substantially increase. In addition, more USATCs and schools could be opened quickly, if required.[17]

It is difficult to estimate the surge capacity of this system.[18] However, an examination of World War II accession rates of enlisted personnel is enlightening. Figure 3.3 shows the expansion rate of enlisted personnel by year during World War II.

As the figure shows, the training system began to increase output in 1940 and trained over one million soldiers that year. In three years, over six million enlisted soldiers were trained for combat. Today these soldiers would not be ready for heavy combat after this training, but they would be ready for advanced training.

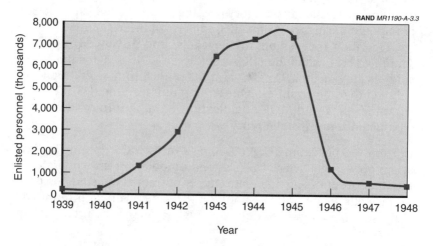

Figure 3.3—World War II Personnel Expansion Rate

---

[17]Fort Benning mobilization plans during the 1980s accounted for an annual infantry OSUT training rate of 78,000 graduates a year. Maximum capacity is even higher if other infantry training sites are opened, such as Fort Polk was during the Vietnam war.

[18]It is difficult to determine the Army's maximum accession rates, partly because TRADOC no longer plans for a full mobilization. The current training base is based on a partial mobilization plan to expand the Army to meet the two-MTW (major theater war) threat. Detailed mobilization planning exists only for a partial mobilization. 1991 was the last year a full mobilization ARPRINT (Army Program for Individual Training) was drafted.

In our analysis, we allow the basic and MOS training capacity to vary from the current four brigades every four months to 20 brigades every two to eight months. Such an output seems well within the Army's capability; many more soldiers received basic and MOS training in World War II.

Noncommissioned officers. Promoting current E-5s up to fill expanded E-6 slots would be difficult because in a typical line squad, there are only two E-5 fire team leaders for every E-6 squad leader and thus not a large pool of promotables. Perhaps more importantly, squad leaders generally mature over the course of several years. These are leadership positions that Army ground forces require in large numbers, and these positions also demand a level of knowledge that cannot be reproduced in a short time span.

Training courses alone cannot produce experienced E-6s, but they can teach some of the most critical skills that will be required. Current peacetime BNCOC courses produce about 576 or 3.6 divisions' worth of E-6 tank commanders per year, about a third of the nine divisions per year required for six NTCs. However, based on past experience, expanding E-6 NCO courses will not be a significant problem.

Enough certified E-6s could be produced today, but because experience is the major requirement for a highly qualified E-6, there would be concern that these NCOs were not as experienced as desired. Yet if combat ready units did not have to deploy immediately after completing their advanced training, more on-the-job training (OJT) could occur, providing additional experience.

Officers. New officers are accessed from one of four sources: Reserve Officer Training Corps (ROTC), the U.S. Military Academy at West Point, Officer Candidate Schools, and direct commissions from either the enlisted force or the civilian pool.

During peacetime, the ROTC camps on college campuses across the country provide most of the new officers. That situation would change during a full mobilization. The ROTC leadership program can be shortened from four years to two years, but this does not represent a significant surge capability. In past national emergencies West Point classes were accelerated, but those efforts are now considered a mistake. West Point's purpose is to produce career officers;

producing new second lieutenants is secondary. Direct commissions accounted for over 100,000 new officers in World War II, but these were primarily noncombat MOSs such as chaplains, lawyers, dentists, or technical and administrative posts and are not a source for O-1s in combat MOSs.

OCS is the only source for officers with surge capability. During a national emergency, Officer Candidate Schools serve as the "throttle" for officer accession, as demonstrated during past wars (see Figure 3.4).

Currently, there is one Federal Army OCS program for infantry officers located at Fort Benning, the Branch Immaterial Officer Candidate Course (BIOCC).[19] BIOCC provides the basic knowledge necessary to become a second lieutenant, regardless of branch of

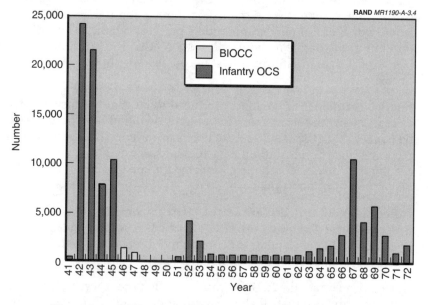

Figure 3.4—OCS and BIOCC Graduation of Infantry Officers

---

[19]Many part-time National Guard OCS programs exist which can be completed on weekends over the course of a year.

service. It currently runs for 14 weeks. After completing it, newly commissioned second lieutenants proceed to their individual branch schools such as the Infantry Officer Basic Course (IOBC) and the Armor Officer Basic Course (AOBC) upon graduation. The IOBC course length is 16 weeks.

Current TRADOC mobilization plans call for the termination of the peacetime BIOCC, after graduating the classes in session, and converting it to an Infantry Officer Candidate Course, sometimes referred to as a Branch Officer Candidate Course (BOCC). BIOCC responsibility would transfer to facilities located at Camp Robinson.[20] The purpose of the infantry BOCC is to turn civilians directly into infantry second lieutenants, effectively combining the tasks of BIOCC and IOBC. This combined training course would run 24–26 weeks.

BIOCC graduation rates in recent years have averaged around 500 per year. If the BIOCC were converted to an infantry BOCC, training output could be increased quickly. With 15 additional staff personnel, training output could be increased to about 2,000 per year. During a full mobilization, more trainers and resources could be added to increase infantry officer production to Vietnam-era rates, about 7,000–8,000 per year.[21] With this rate of production, the requirement of about 918 to 1,836 infantry O-1s per year could easily be met.[22]

Armor officers follow the more usual officer accession plan. First, they are trained at a BIOCC to receive their commissions, then they are sent to the Armor Officer Branch Course (AOBC) for specific MOS training. The current AOBC is capable of running an abbreviated course of 79 days (11.3 weeks), running four classes of 64 officers each concurrently and, after a staggered start, graduating roughly 768 armor officers per year. If only combat-essential instruction is

---

[20]Of all the branches, only the infantry, field artillery, and engineer branch schools are required by TRADOC's mobilization plans to maintain a current capability to run a BOCC.

[21]Based on comments by COL Scott Armbrister at OCS, June 3, 1997.

[22]The assumption here is that all expanded light divisions are regular infantry. Additional officer training beyond BOCC—such as airborne, ranger, or special forces training—would be ideal for at least some of the new infantry officers.

performed, the class length can be cut down to 60 days, graduating 24 classes or about 1,536 officers per year.[23]   Our calculated requirement for armor officers is about 738 to 1,476 O-1s per year and thus armored officer accession should not be a constraint, given a full mobilization.

EQUIPMENT PRODUCTION

Currently there is not enough modern equipment to outfit 40 divisions of soldiers; more would have to be produced. How much and what type of new equipment must be produced, and how rapidly could this new equipment be produced?

Division Sets Required

The amount of new equipment that must be produced to outfit 40 divisions is the difference between the total equipment required for 40 divisions and the current inventory of equipment, both expressed in heavy-division-equivalents (HDEs) of equipment. The first problem is to determine how much current equipment exists.

Current Big Five inventory. Examining all Army divisional equipment would be a daunting task. To simplify our problem, we focus only on the major combat equipment items planned during the 1970s and acquired during the 1980s and 1990s, ignoring other key division equipment assets.[24] The so-called Big Five consists of the

---

[23]Telephone interview with COL Douglas Slater, Commander Second Squadron, 16th Cavalry, Fort Knox, KY.

[24]We did take a brief look at the truck industrial base to confirm that expansion of the Army's truck fleet would not be a bottleneck. We looked at production rates, lagtimes, and ramptimes for the 2.5- and 5-ton medium tactical vehicles (MTVs), the Oshkosh 10-ton HEMTT trucks, and the Oshkosh HETs. The industrial base for trucks differs from the other five combat system industrial bases in two important respects: it is "hot" and it is much more "commercial" in nature. First, it is "hot" in the sense that brand new vehicles are running off the assembly lines today (the HET trailers are probably the most critical for expansion, since there is no close commercial counterpart and there is only one manufacturer). Second, DoD leverages off the truck commercial base to a far greater extent than it does for combat systems. For example, there are no major components for the MTVs that are specifically military in nature—in other words, there are commercial equivalents for all components of the MTV variants. In a national emergency, the government could confiscate all commercial engines and adapt them for MTVs. Finally, the Army could alleviate expansion pains

Abrams Main Battle Tank, the Bradley IFV, the Apache helicopter, the MLRS launcher, and the Patriot launcher. These equipment items form the core of the Army's current combat power. Any major Army expansion for years to come would require an expansion of these items.[25]

Abrams main battle tank

In June 1973, contracts were let to build prototypes of a new tank designated the M-1 and later named the Abrams. The prototypes were delivered in 1976. Production at the Army Modification Center in Lima, Ohio, began in 1979, with the first production M-1 delivered to the Army in 1980. There were 3,268 M-1s produced before this model went out of production in 1984.

Beginning in 1985, the M-1A1 model was introduced, and a total of 4,771 have been built for the Army.[26] Then the M-1A2 was introduced, but declining Army procurement budgets and force structure have reduced the scope of the M-1A2 program. To date, only 62 new M-1A2s have been produced. Instead of new production, the M-1A2 program has become an upgrade program, with selected M-1s being overhauled and upgraded to the M-1A2, which will keep the Lima facility operating.[27] The total number of all M-1 models in the current Army inventory is about 8,101.[28] Figure 3.5 shows the production and conversion of the various M-1 models to date.

---

to some extent by leasing commercial transport vehicles like it did in the Persian Gulf in 1991. (Data provided in part by Mr. Dennis E. Mazurek, Deputy Project Manager for medium tactical vehicles, PEO-Ground Combat and Support Systems.)

[25]We are not considering current new weapons systems such as the Comanche and the Crusader, which would probably also be produced. Neither are we explicitly considering the full array of electronics upgrades upon which the modernized Force XXI depends. We will examine these force improvements only parametrically.

[26]Other M-1 versions have been built for the Marines and for foreign military sales abroad.

[27]The M-1A2 upgrade program (120 M-1A2s each year) stabilizes the supplier base for the Abrams tank. General Dynamics Land Systems (GDLS) Division is also refurbishing M-1A1 Abrams at Lima and at Anniston Army Depot under the Abrams Integrated Management for the 21st Century (AIM XXI) program, which will allow the Army to keep tanks in service for up to 40 years.

[28]Some M-1s have been lost through accidents over the years; others are undergoing major repairs. These losses are relatively small and will be ignored because we do not have good data for these losses for all weapon systems.

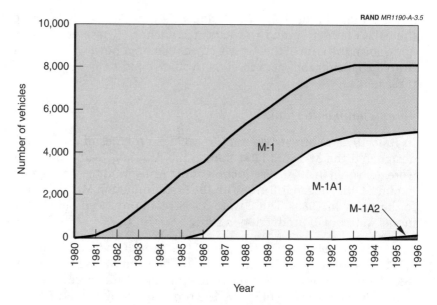

Figure 3.5—Abrams Production

It is difficult to predict when the Army will move to the production of a new-generation tank. Currently the Abrams is not funded past 2003. One alternative is to produce an Abrams upgrade (M-1A4), which could improve combat effectiveness and ensure a "warm" tank industrial base until the Future Combat System (FCS) is ready for production. Another alternative is to funnel all future R&D funds to the FCS so it can be fielded earlier than the scheduled 2015 date.[29]

Bradley

The Bradley was born with the idea of extending the capability of troop carriers beyond that of armored personnel carriers such as the M-113, in order to provide a dedicated main armament that could support the mounted infantry squad and defeat enemy light armor. Prototypes of the M-2 and M-3 were built in 1975, followed by another series of prototypes in 1978. The first production vehicles

---

[29]CPT Todd Tolson, "Building Tanks at Lima," *Armor*, November–December 1996.

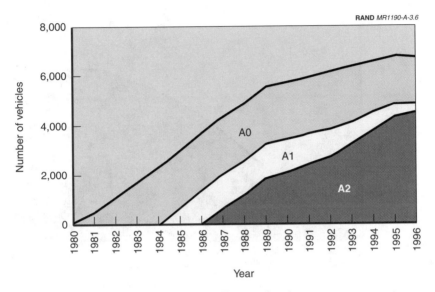

RAND *MR1190-A-3.6*

Figure 3.6—Bradley Production

were delivered in 1981 and production was completed in 1994, with about 6,720 vehicles delivered by then.

The first improved version of the Bradley, designated the M-2/3A1, was introduced in the beginning of 1986. In 1987, the A2 upgrade was introduced; in 1997, the A3 version. Conversion of earlier versions of the M-2/3 to the M-2/3A3 is under way at a rate of about 216 vehicles per year, but no new production is expected, leaving current inventory at about 6,720. Figure 3.6 shows the production rate and the conversion rate of the different M-2/3 versions.

Apache

Drawing on the Vietnam experience but focusing on the confrontation with the Soviet Union in Europe, the Army established goals for a new helicopter design. With the anti-tank mission as primary, the Army asked for designs in 1972. By 1973, competitive prototypes were authorized to be built by Hughes and Bell Helicopter, with the Hughes prototype emerging as the winner in 1976. The first produc-

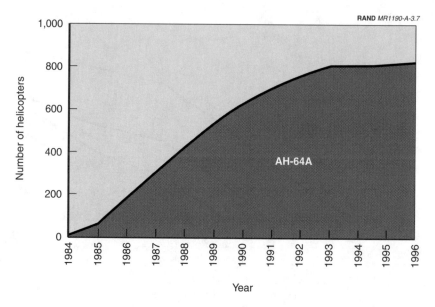

Figure 3.7—Apache Production

tion AH-64A was delivered to the Army in 1983. By 1996, 811 airframes had been produced for the Army (see Figure 3.7).[30]

Beginning in 1997, the Army decided to upgrade all these AH-64s to Longbow AH-64Ds.[31] This conversion program is planned to continue until most of the AH-64As have been converted.

MLRS

The MLRS concept began with the formulation of a requirement for a rocket system with high fire rates to supplement conventional artillery tubes, especially in a counterbattery mode. The ability to surge the volume of counterbattery fire was the primary design objective. In 1977, the Army awarded competitive contracts to Boeing

---

[30]Sales to the U.S. Army have been completed, but sales abroad continue.

[31]The Longbow weapon system consists of a modified AH-64A airframe, a Fire Control Radar mission kit, and a Longbow Hellfire missile. Other changes include increased electrical power, expanded forward avionics bays, increased cooling, upgraded processors, MANPRINT crew station, and 701C engines.

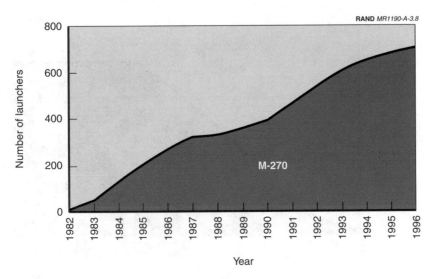

Figure 3.8—MLRS Production

Aerospace and LTV's Vought Corp. The first MLRS production models were delivered to the Army in 1982. To date, about 794 MLRS launchers have been produced for the Army and ARNG (see Figure 3.8).

In 1997, the Army began a retrofit program to modify its current M-270 launchers to an M-270A1 model with an improved fire-control system and an improved hydraulic system for better slewing, elevation, and overall system responsiveness.

Patriot

The Patriot missile system began as the SAM-D program in 1964. After several years of political controversy over its requirements and capabilities, the SAM-D entered full-scale development in 1976 and was renamed the Patriot.

A Patriot fire unit consists of a radar set, the engagement control station (ECS), the equipment power plant (EPP), an antenna mast group (AMG), and eight remotely controlled launchers. To date there have been 94 fire units produced (see Figure 3.9).

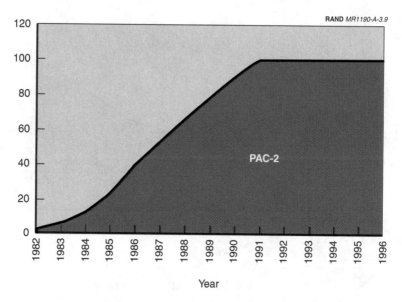

Figure 3.9—Patriot Production

Several upgrades of Patriot have been made that have involved changes in the radar software (PAC-1) and improvements to the missile's warhead and fuze (PAC-2). The PAC-3 involves additional modifications to the radar software and the missile warhead and fuze, but no new fire units will be added.

Equipment as division sets. How many division sets are represented by this equipment?

Abrams

An armored division has six armored battalions with 58 tanks per battalion for a total of about 348 tanks; a mechanized division has five armored battalions for a total of about 290 tanks.[32] To calculate the number of tanks in a nominal HDE, we take a weighted average of these two numbers to reach an estimate of about 309 tanks per

---

[32]This study is based on the division design in place in 1997.

HDE.[33] If we then divide the total number of M-1s (8,101) by 309 M-1s per HDE, we conclude that there are about 26 HDEs of M-1s.

Bradley

An armored division has four mechanized battalions for a total of about 240 Bradleys, and a mechanized division has five mechanized battalions for a total of about 300 Bradleys. Taken together, a nominal Bradley HDE could have the weighted average of these two division types, about 280 Bradleys. If we divide the 6,720 total Bradleys by 280 Bradleys per HDE, we get about 24 HDEs of Bradleys, close to the number for M-1s.

Other Big Five

Estimating the number of M-1s and M-2/3s per HDE is straightforward because these are both division-level assets. However, estimating HDEs of Apaches, MLRS, and Patriot is more difficult because these are also held at corps and EAC levels. For example, about 22 to 44 Apaches are held as division assets, and about 44 more Apaches are held as corps assets. If we assign one-third of the corps assets to each division for counting purposes, then a nominal HDE consists of about 35 to 55 Apaches.

However, air assault, airborne, and light divisions also require attack helicopters. If the Army expanded its heavy force structure in the future, we do not know how many light forces would also be expanded. In the Gulf War, the 101st Air Assault Division was particularly effective in the battle against the Iraqi armor units. Thus, it is difficult to know the number of Apaches required in a heavy force expansion. The same assignment difficulties hold for MLRS and for Patriot.[34]

Therefore, we decided to use the following approach to estimate the number of equipment items per HDE. To do this, we define HDEs of Bradleys, Apaches, MLRS, and Patriots based on the assumption that

---

[33]We use a two-to-one weighting average because the Army has tended to balance two mechanized divisions for every armored division.

[34]For example, Patriot battalions can have anywhere between three and six fire units. They are also only assigned to corps and echelon above corps commands.

the current force structure is balanced. If there are about 26 HDEs of M-1s, then we assume that the total inventory of Bradleys, Apaches, MLRS, and Patriots also amounts to about 26 HDEs. Then we can assign a number of Bradleys, Apaches, MLRS, and Patriots to a single HDE.[35] Table 3.3 shows this assignment.

Expansion Capability

If there are currently about 26 HDE sets of equipment, then to expand to 40 HDEs, 14 additional HDEs of equipment would have to be produced. Let us examine this equipment expansion process.

Table 3.3

Nominal HDE Equipment

Weapon Systems	Total Production	Number per HDE	HDEs
Abrams tank	8,101	309	26
Bradley IFV	6,720	258	26
Apache helicopter	821	32	26
MLRS launcher	794	30	26
Patriot FU	101	3.8	26

---

[35] At the time our exploratory analysis was conducted (1997), Army divisions were structured under the Army of Excellence (AOE) design. The new Army XXI Heavy Division Redesign that was announced in June 1998 is now currently under way and should be completed by the end of 2000. According to Directorate of Combat Developments, U.S. Army Infantry Center, the primary change is to drop a company from every battalion. This means the total tanks per battalion will drop from 58 to 45 and total Bradleys per battalion will drop from 60 to 54. Accounting for the Force XXI redesign would decrease our nominal HDE definition for Abrams—from 309 to 240— and increase by 30 percent the assumed initial inventory of Abrams—33 divisions instead of 26. A lower HDE value for Force XXI would increase the production rates used for our analysis, *ceteris paribus*. The parameters of the exploratory model outlined in Appendix B could be calibrated to reflect these changes easily enough (specifically, the InitEqp, InitProdRate, and MaxProdRate parameter values), and further exploratory analysis would determine the ramifications. However, we should note that the relative differences in rates among the Big Five—as well as the ramptimes and lagtimes—would remain the same. In all likelihood, the conclusion that brigade training centers are the main bottleneck to acquiring 40 heavy divisions would remain the same.

Assessing the capacity of the industrial base is beyond the scope of this analysis; an extensive effort would require looking at the prime contractor base, the supplier and vendor base, and the depot industrial base of government-owned maintenance and repair facilities. Here we have focused on the production capability of the prime contractor facilities and manufacturing resources. Most of our industry parameters are estimates that were gathered through interviews with various production managers, Army program managers, military system coordinators, acquisition experts, contractors, and suppliers.

All of the Big Five equipment items have been upgraded during their lifetimes and are currently being upgraded to newer models in an existing plant. During an expansion, all five weapon systems could be upgraded while facilities for new production are being prepared. Whether modernization could continue without slowing down expansion rates during the production of new units depends on the nature of the weapon system industrial base. We assume that the Army would want to produce new equipment items as quickly as possible rather than delay or slow their production by trying to simultaneously upgrade old equipment.

The current state of the industrial base for the five weapon systems can be described as "warm"—that is, plants, tooling, and labor are currently being used to upgrade older models to newer models in all five cases. However, no units are being produced from scratch for the most part (other than a trickle of foreign military sales in some cases). Because the industrial base is not producing new units (it is not "hot"), suppliers of key components are not always available. For example, the supplier of the Abrams tank engine, Allied Signal, would need at least two years to begin providing tank engines to the Lima facility.[36] This translates into a startup delay or lead time to produce new tanks.

---

[36]Most of our Abrams data come from Lou Lypeckyj and Don Livingston at the Abrams Program Office, Acquisition and Production. They note three explicit assumptions underlying the figures used in this report. The first is that an engine vendor remains available. The current supplier, Allied Signal, has already closed one of its facilities (because it no longer is producing engines for the Abrams program). The second assumption is that armor vendors remain available (which is classified). The third assumption is that ballistic welders can be trained during the two-year lead

Startup delays and production rates vary across the five weapon systems. Below we discuss each weapon in more detail. Table 3.4 shows a summary of the production parameters.

In addition to a startup delay while crucial suppliers and vendors are mobilized to provide subcomponents, there is a ramp-up phase of production. Reaching maximum production rates using a full schedule of 2.5 to 3 shifts takes time. Even with a mature design, there are always design problems and startup problems, so a limited production release program is always necessary and a good idea. Production managers want to maximize production but also mitigate risk; they never want to commit too much before they find flaws in the design. Ramp-up rates and ramptime (the length of time to reach maximum production) also vary across weapon systems.[37] Ramp-up manufacturing rates in Table 3.5 are based on a two-year lead time.[38]

Abrams. Currently, and for the next several years, 120 M-1s per year are being upgraded to the M-1A2 model at the Lima plant. If Lima changed to producing new tanks, its rate of production and startup delay would depend on whether the production base was "warm" or "cold."

A "warm" production base exists if the plant has been out of operation for less than four months. If so, Lima would have about a two-year lead time before building completely new tanks because it would require about that long to acquire a new engine base (i.e., reconstitute the vendor base, etc.). On the other hand, if the production base is "cold," that is, the plant has been out of operation for

---

time delay (which is possible but tight, given that the training program takes 18 months).

[37]We gathered detailed ramp-up data from the following helpful sources: LTC Rick Ryles (Apache Program Manager's office, Redstone Arsenal, Huntsville, Alabama); Dave Parabek, the Production Manager for PM Bradley, U.S. Army Tank Automotive and Armaments Command; Dave Hagarty (manufacturing program manager for the Patriot and Hawk, office of the Andover plant manager); Dennis Vaugh, MLRS Program Office; and COL Steve Kratter, M270A1 Remanufacturing Program.

[38]In most cases, production managers said they would take advantage of the lead time to train required specialists and add additional tooling in anticipation of receiving vendor and supplier components toward the end of the lead time. For example, the Abrams ramp rate of 450 tanks for a ramp length of one year is based on already having two years to prepare.

Table 3.4

Production Summary

Weapon System	Startup Delay or Lead Time to Produce New Units (years)	Production Rates per Year for 1/2.5 shifts	Maximum Production Rate (in HDEs/year)	Lead time to Build 2nd Plant (years)
Abrams tank	2	360/900	2.91	2
Bradley IFV	2	240/600	2.33	3
Apache helicopter	2.5	72/144	4.65	1
MLRS launcher	2	48/120	4.00	3
Patriot radar	2.5	12/24	6.32	3

Table 3.5

Manufacturing Ramp Rates

Weapon System	Production Rates per Year for 1/2.5 Shifts	Maximum Production Rate (in HDEs/year)	Ramp Time After Lead Time (years)	Ramp Rates as Percentage of Max-Prod-Rate (average for the year)		
				1st year	2nd year	3rd year
Abrams tank	360/900	2.91	1	50%	100%	100%
Bradley IFV	240/600	2.33	1	50%	100%	100%
Apache helicopter	72/144	4.65	1.5	50%	75%	100%
MLRS launcher	48/120	4.00	1	50%	100%	100%
Patriot radar	12/24	6.32	0.5	50%	100%	100%

more than four to six months, Lima would have about a five-year lead time to begin building new tanks because of vendor mortality. Key subcomponent manufacturers such as the tank engine contractor and depleted uranium armor suppliers would have to reenter the industry and rebuild their facilities. Critical labor skills such as ballistic welding and numerical machine operators are difficult to find, and training ballistic welders can take 18 months.

Rebuilding a second plant, like the old Detroit tank plant, would take about two years to build the actual physical structure. During this two-year construction period, a warm industrial base could be made ready to produce new tanks. A cold industrial base would require a total of five years before all necessary components would be available from suppliers.

After the startup delay, if the Lima plant maximized its production by working 2.5 shifts per day, it could produce about 900 tanks per year, or about 2.9 HDEs per year. A second plant could increase that output to about 1,440 tanks per year.

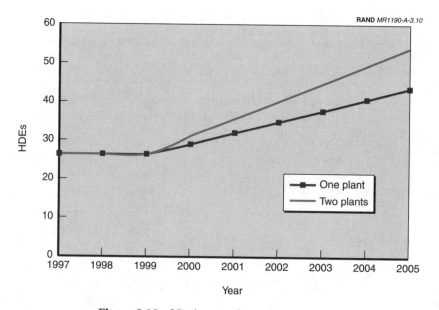

Figure 3.10—Maximum Abrams Production

At these production rates, one plant working 2.5 shifts would require about six years, including startup time, to produce 14 more HDEs of tanks. If a second plant were opened, also working 2.5 shifts, the same number of tanks could be produced in about four years.

Figure 3.10 shows the M-1 expansion rate for one and two plants using nominal values.

Bradley. Bradley production began in 1980 at the San Jose plant. From 1980 to 1982, the production rate of the initial models (A0s) was 50 per month using two shifts. When the A1 and A2 models went into production at the San Jose plant, the production rate of these models was 35 per month while working one shift, and 65 to 70 per month while working two shifts; a surge capability was achieved of 90 per month while working 2.5 to 3 shifts. In 1997, the main production plant moved to York, Pennsylvania, where the production rate is now planned to be about 20 per month for one shift, 40 per month for two shifts, and 50 per month for 2.5 to 3 shifts.

The York plant is currently upgrading the A1 and A2 models to the A3 model. Producing new vehicles again would require a lead time of about 24 months, primarily because an engine component supplier would need to be found. Acquiring this component from a subcontractor is estimated to take about 24 months.

If a second production facility were required, building a new plant would take about 36 months before new production could begin. The constraint is not the process of building a facility; it only takes three to six months to construct a building. The constraint is the duplication of necessary plant equipment and machine tooling.

With these nominal parameter values, 14 additional HDEs of Bradleys could be built in about six years with one plant working 2.5 shifts. If another plant were opened, 14 HDEs of Bradleys could be produced in about four years (see Figure 3.11).[39]

_____

[39]Most of the production data and much of the industrial capability information was provided by Dave Parabek, Production Manager for PM Bradley, Acquisition Division Office, U.S. Army Tank Automotive and Armaments Command, Warren, Michigan.

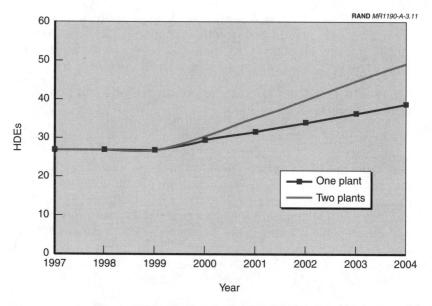

Figure 3.11—Maximum Bradley Production

Apache. The only McDonnell Douglas Helicopter Systems (MDHS) Apache assembly plant is located in Mesa, Arizona, and was built in 1981. Past peak production of the first Apache model, the AH-64A, at this plant was 12 per month working two shifts. Since the early 1980s, however, the capacity has declined due to a loss of some tooling. The maximum surge rate today given existing tooling is about 8 Apaches per month. If additional tooling were provided, the existing facilities could again produce 12 Apaches per month.

The lead time to produce a new Apache is 2.5 years. This is the period required to mobilize the supplier base to provide the necessary components. If another assembly plant were to be built, it would probably require about 4 years before it could be ready. Given these nominal parameter values, it would require about 5 years and 4 years, respectively, for one and two plants to build 14 more HDEs of Apaches, as shown in Figure 3.12.[40]

---

[40]Most of the production history and industry data for this section was provided by Victor Burgos and LTC Larry Thomas, Apache Project Managers Office, Aviation and Troop Support Command, and Hal Klopper, MDHS Public Relations Office.

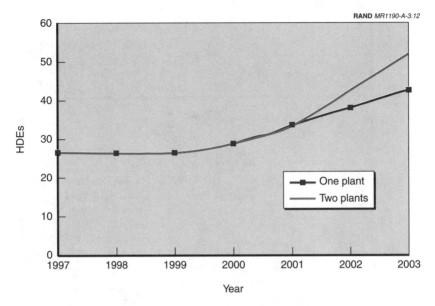

Figure 3.12—Maximum Apache Production

MLRS. The main MLRS plant is run by Lockheed Martin Vought Systems (LMVS) at Camden, Arizona. Camden MLRS launcher manufacturing is a combination of manufacturing and assembly. LMVS vendors supply components for final assembly. Of the major structural components, the hull is supplied by the Bradley plant at York, while the base cage and turret are built by LMVS. The production rates for new M270 launchers are about 4 per month for one shift and 10 per month for 2.5 shifts.

The lead time to produce new launchers is 2.5 years from a cold production base, with the production of the base cage and turret as the biggest constraint. Construction of a second facility with required reinforced walls and a roof designed to lift in case of explosions would take an estimated three to five years.

A small number of MLRS systems are currently being produced at the Camden plant for the ARNG and foreign buyers. Although 20 launchers are expected to be built in 1997, a contract lead time of two years for vendor components can be assumed. Assuming 30 launchers per HDE, four HDEs of MLRS launchers could be produced each

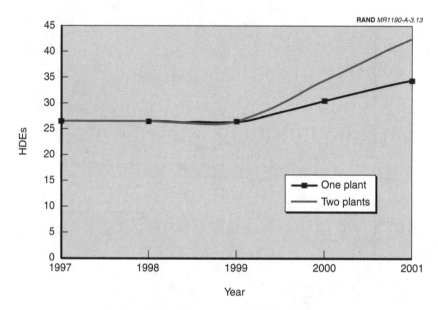

Figure 3.13—Maximum MLRS Production

year after the initial two-year delay. Two plants would double that rate to eight HDEs per year (see Figure 3.13).[41]

Patriot. The Patriot contractor is Raytheon, whose main assembly plant is located in Andover, Massachusetts. In a Patriot fire unit, the radar requires the longest time to produce, leading to a maximum production rate of 12 fire units per year, after startup, increasing to 24 fire units per year.

Currently the Andover plant is upgrading the PAC-2 radar. The startup delay or lead time necessary before Andover could produce completely new Patriot fire units is about 2.5 years. Building a second plant would require about 3 years. Again, the time constraint here is primarily imposed by the much-reduced supplier base.[42]

---

[41]Much of the MLRS information provided here was obtained through interviews with Dennis Vaugh, MLRS Program Office, and Colonel Kratter, M270A1 Remanufacturing Program.

[42]The expansion lead time also depends on the time to build transformers and to train radar system engineers.

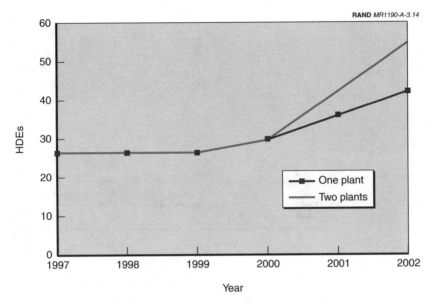

Figure 3.14—Maximum Patriot Production

Patriot production rates are mainly constrained by the availability of electronic testing equipment. The Andover plant is capable of producing new fire units and modernizing PAC-2s to PAC-3s at the same time but at a slower rate than if production or modernization were done alone.

When the production rates with nominal values are plotted, Figure 3.14 shows the rate of production of new Patriot fire units for one and two plants. Fourteen HDEs of fire units could be produced in about four years and five years in one plant and two plants, respectively.

MAXIMUM EXPANSION RATES TODAY

Expanding Personnel

Figure 3.15 compares the potential training capacity of the advanced training system with the potential availability of units ready for advanced training.

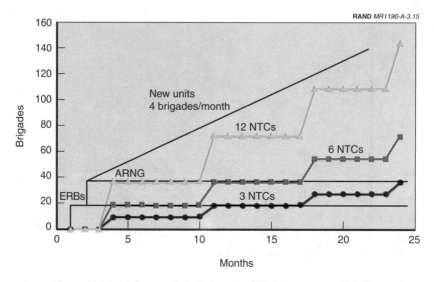

Figure 3.15—Advanced Training Capacity Versus Availability

Here we have plotted three different advanced training capacities: for 3, 6, and 12 NTCs. In all cases, there is a lag of three months before the sites are available. Then we have plotted the potential availability for advanced training of forces from the ERBs, the ARNG and newly formed units (at a rate of four brigades per month).[43]

Comparing the potential capacity of the advanced training system with the potential availability of units ready for such training, we see that with nominal values, there are likely to be more units ready for advanced training than can be accommodated. However, varying some of the parameter values shows that the bottlenecks in the training could change. For example, if neither ARNG forces nor newly formed units were ready for advanced training until nine months and 12 NTCs were operating at three months, then the advanced training cumulative (or NTC training demand) capacity would exceed the potential supply of trainees.

---

[43]Four brigades a month is illustrative. The current rate is about three brigades per month, but during Vietnam about 15 infantry brigades per month were trained.

Because the values of these parameters cannot be known with precision today and even less in the future, our exploratory analysis below examines the effect of varying these parameters through a range of values to examine their effect on the output of trained units.

Expanding Equipment

Figure 3.16 compares these data for the production of different weapons systems from one plant. With one plant working 2.5 shifts for each equipment item, about seven years would be required to produce 14 new HDE sets of equipment for M-1s and M-2/3s. These items are actually division holdings, and we assume that their production rate would determine the overall production rate of new divisions.

If two plants were opened for all systems, then 14 new HDEs of M-1s and M-2/3s could be produced in about five years with a normalized number of Apaches, MLRS, and Patriot even faster.

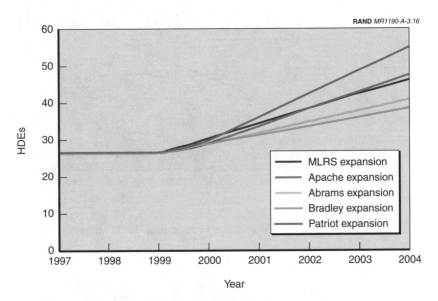

Figure 3.16—Maximum Production: One Plant

Maximum Expansion Rates

Combining the expansion rate for personnel and equipment provides an estimate of the overall maximum expansion rate. In Figure 3.18, we have combined the maximum personnel training rates[44] (see Figure 3.15) with the maximum equipment production rates, using the M-1 and M-2/3 as the standard (see Figures 3.16 and 3.17).

If six NTCs were operating, then during the first two years after full mobilization, about 26 HDEs could be trained and equipped with current equipment holdings. At about the same time, new equipment would begin to be produced, although not as fast as personnel could continue to be trained. Indeed, 40 HDEs of trained personnel

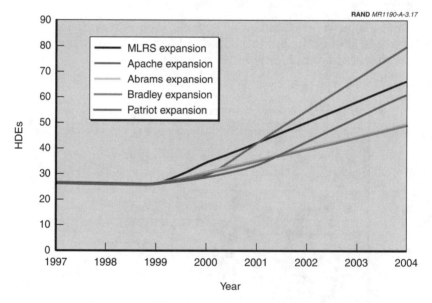

Figure 3.17—Maximum Production: Two Plants

---

[44]Actually, Figure 3.18 shows the time at which an advanced training facility is ready for a unit to enter to begin training. We have plotted the time at which a unit has completed its training. Furthermore, we have averaged the process to produce a smooth curve.

could be produced by about four years. To equip these 40 HDEs of personnel would require about seven years for one operating plant for each equipment item and about five years for two operating plants.

These estimates were derived by using nominal parameter values in a simple flow model. These are our best estimates, but it is difficult to precisely estimate these parameters even today. The further in the future one goes, the more difficult the estimation becomes. However, our goal in this research is not to provide an accurate point estimate of the expansion rate of the current system, but rather to identify which factors could restrict the system's expansion in the future. For this reason, we will use an exploratory analysis below to examine the expansion process over a range of parameter values to determine the relative effects of each factor.

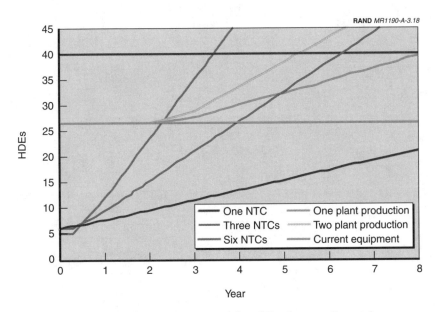

Figure 3.18—Maximum Personnel and Equipment Expansion

Chapter Four

EXPANDING THE ARMY'S HEAVY BRIGADES IN THE FUTURE

THE MODEL

Figure 4.1 is a flow diagram of the basic parametric model used in the explorations of heavy force expansion timing in the future. The parameters used, their nominal values, and the high and low values of their ranges are detailed in Appendix B.

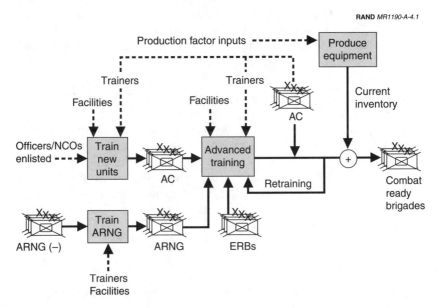

Figure 4.1—Flow of Parametric Model

DOCUMENTING AN EXPLORATORY ANALYSIS

There is an inherent difficulty in documenting an exploratory analysis that stems from the intensely interactive nature of the explorations. As in a scientific laboratory, one can approach an exploratory model with a hypothesis to be tested and a careful plan for that testing. Some of the explorations were, indeed, carried out in that fashion. On the other hand, the model parameters appropriate for describing the expansion system up to 20 years in the future are quite speculative at this point. This prompted a much more modest goal from the research than is typical in the laboratory. We were testing the common wisdom on expansion by primarily looking for surprises in the model outcomes throughout the plausible ranges of the parameters. This makes the research much more of an exploration or "wandering around" looking for something interesting.

To document this wandering around, we have chosen a two-part format. The first part will be what was actually observed in the explorations of the model that were interesting enough to note. To the extent possible, these will include quantitative descriptions. The second part will be the interpretation of those observations (possibly including an interpretation of things that were *not* notable). The interpretation will translate the observations into what can be said about expansion of the Army in the future.

BOTTLENECKS IN EXPANDING HEAVY BRIGADES

From Figure 3.18 in Chapter Three, it is a clear that the primary bottleneck today for heavy force expansion is the number of advanced training sites or NTCs. This fact provides a useful means for discussing our explorations of the outcome space of the model. By starting with the training bottleneck, the exploration then becomes one of exploring and documenting what, when, and how other elements become bottlenecks instead.

One NTC Only

Figure 4.2 shows how fast the Army could expand from today if it used only the National Training Center as its brigade-level training facility. If the system is never allowed to have more than one NTC,

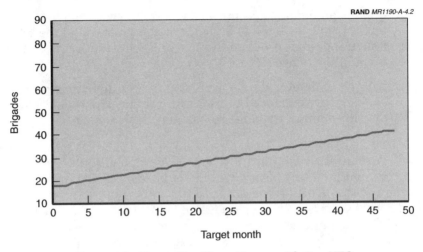

Figure 4.2—Expanding Heavy Forces with One NTC

changing the other parameters has little effect, and out to four years, the bottleneck in expanding the Army is brigade-level training.[1] In sensitivity analysis terms, Figure 4.2 is insensitive to most of the parameters in the exploratory model. In exploratory modeling terms, moving the slider bars for most of the other model parameters does little to change the display.

Observations

- The one parameter that most affects Figure 4.2 is the number of initial ready brigades. An increase (or decrease) in ready brigades to begin with produces basically the same increase (or decrease) in generated brigades at any later point.

- If the advanced training time is reduced, the slope of Figure 4.2 increases. For example, if the training time is cut in half, the slope doubles, producing about 60 brigades in four years.

- If heavy brigades need retraining in less than four years, that will "clog" the pipe and cause the system to spend all its time doing retraining once retraining begins.

---

[1]Where, as earlier, this includes division-level training as well.

- Initial equipment can cause a slowdown in generated brigades, but only below about 30 brigades of equipment (and then not much—nominal production rates are about the same as the training rate of one NTC).

- In order to depress the slope of Figure 4.2 further, there have to be few or no enhanced brigades or National Guard and fewer than the nominal training sites for individual training.

Interpretation. With only one NTC, the training bottleneck is so severe that the Army basically needs to have its capabilities in the ready force. Cutting the advanced training time (e.g., through advanced training technologies) would help, but it would have to be cut in half to produce even 50 percent more total brigades. If retraining of brigades is required during the expansion, the point at which it begins is the last point at which new brigades will have been produced. At least half the current equipment would have to become obsolete in some way before the industrial base came into play as a bottleneck. Even then, it would depress the generated brigades only slightly over the rate in Figure 4.2. Only drastic reductions in enhanced brigades, National Guard, and individual training facilities would cause further slowdown.

Two NTCs

Figure 4.3 shows how fast the Army could expand from today if it built and manned a second combat training center (in three months). The rate of expansion is roughly twice that of Figure 4.2 and produces about 60 brigades in four years. The expansion rate is dominated again by the advanced training requirements. The sensitivity of this rate to most other parameters, while somewhat larger than that for one NTC, is still low.

Observations

- The largest effect on Figure 4.3 is, again, the number of initial ready brigades. An increase (or decrease) in ready brigades at the beginning of the expansion produces generally the same

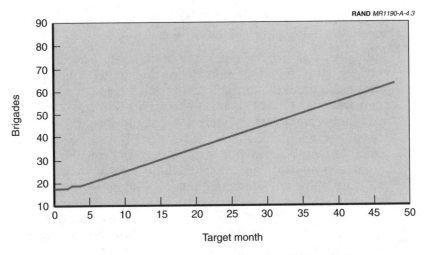

RAND *MR1190-A-4.3*

Figure 4.3—Expanding Heavy Forces with Two NTCs

increase (or decrease) in generated brigades at any later point.[2] If there were as many as 45 initial ready brigades, the industrial base would come into play (i.e., at nominal production rates, we would run out of equipment) near the end of the fourth year.

- If the time required for NTC training can be reduced (or increased), the slope of the expansion increases (or flattens). If this advanced training time could be decreased from 2 months to 1.5 months, the industrial base (at nominal production rates) would come into play.

- If production rates were upped to two plants and 2.5 shifts each, the production of equipment could stay ahead of the advanced training bottleneck. This, however, assumes having 78 brigades of equipment at the start of the expansion. With the nominal delay of two years in order to get new plants on line and mobilize the supplier base, there would have to be at least 43 brigades of

_____

[2]It could actually produce slightly more. Having more ready brigades initially reduces the need to prepare new training sites. New training sites can diminish the ready brigades in order to provide trainers for those sites.

equipment to keep the pipeline at full production (i.e., to keep there from being a "dip" in the curve in Figure 4.3).

- Individual training can become a bottleneck. The individual training site at Fort Benning currently trains roughly four brigades at a time. If it trained only one brigade at a time and there weren't a total of about 33 ERBs and NG brigades, individual training would affect Figure 4.3 at the four-year mark. Any fewer reserve brigades affects the curve in Figure 4.3 earlier. The effects are similar if two brigades are given individual training at a time and there are not a total of about 24 reserve brigades; if three are trained at a time and there are not at least 15 reserve brigades; and if four are trained at a time and there are not at least nine reserve brigades. Beyond four at a time there is no further effect. That is, there must be at least nine reserve brigades in order to keep the pipeline full, no matter how many individual training sites there are (there is that much delay in getting civilians ready for advanced training).

Interpretation. The situation with two NTCs is more complicated than with one. With a combat training center in addition to the National Training Center, about 60 heavy brigades could be produced and trained in four years (given today's basic parameters). The industrial base can be brought into play more easily, although the changes required (decreasing the advanced training time, having a significant fraction of the equipment base become obsolete, etc.) seem slightly implausible. On the other hand, if two plants were opened, the industrial base capacity could stay ahead of the advanced training output.

The ability to recruit people into the system can be a problem (if conscription is not an option), but only if there is a drop in the total reserve strength.

Three NTCs

Figure 4.4 shows how fast the Army could expand from today if it built and manned two combat training centers (both in three months) in addition to the NTC. This is a particularly interesting

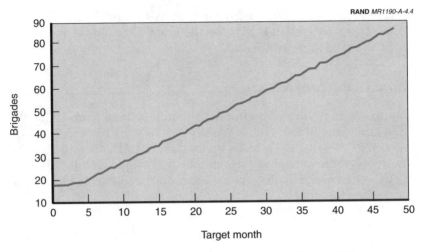

Figure 4.4—Expanding Heavy Forces with Three NTCs

case because of earlier research that indicated the Army could currently man up to three NTCs from the TOE Army.[3]

Observations

- Figure 4.4 is quite sensitive to a wide variety of parameters. There is a drop in the total output of the system if there is *any* drop in the individual training capacity, the advanced training time, the individual training duration, the initial number of ready brigades, ERBs or NG brigades, the initial equipment, the maximum production rate, or even the lagtime in bringing production facilities on line.

- Maximum equipment production rates can stay ahead of the training output from three NTCs.

Interpretation. The situation with three NTCs and today's nominal parameters is "optimized." That is, any future change that would slow down any aspect of expansion would also slow the total system

---

[3]Lippiatt et al. (1996).

output at four years (and often sooner). With three NTCs, then, the total expansion system is right on the knife-edge of maximum production over a four-year period.[4]

More Than Three NTCs

Figure 4.5 shows how fast the Army could expand from today if it built and manned four NTCs. While previous work has suggested the possibility of going to six NTCs on land that TRADOC controls—and there is land available to the military that would handle up to 12 NTCs—the cases above three NTCs can all be addressed with the single case for four NTCs.

Observations

- Individual training can dominate the early expansion rate in this case. If only one brigade at a time gets individual training, the expansion rate becomes limited at about 27 months. If two are trained simultaneously, the rate becomes limited at about 30 months, and for three it becomes limited at about 35 months. Beyond that, the industrial base is the rate limiting factor (visible as a "knee" in the curve of Figure 4.5 for a one-plant/one-shift production rate).

- The maximum equipment production rate (two plants, 2.5 shifts) can keep up with four NTCs, but it falls behind the training rate for five or more NTCs.

- Figure 4.5 assumes an initial equipment stock of 78 brigades. There must be at least 68 brigades of equipment for there not to be a "dip" in Figure 4.5.

Interpretation. It is possible for individual training to be a bottleneck when there are four or more NTCs. As it is relatively easy and inexpensive to increase the number of individual training sites, this

---

[4]This is a surprising result. Because of the discrete nature of many of the parameters (it is difficult, for example, to have 1.5 plants, or 2.2 combat training centers, or 18.6 initial ready brigades), it would be surprising to find another target brigade and target month pair whose achievement would be so sensitive to so many of the system parameters.

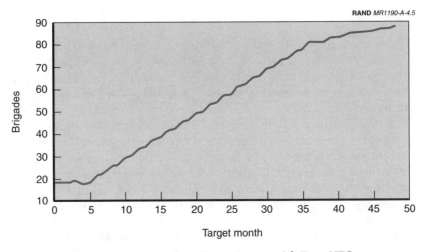

Figure 4.5—Expanding Heavy Forces with Four NTCs

should not be a problem in the future *unless* this is viewed as a surrogate for the ability to attract new recruits. Said another way, if the Army has difficulty attracting civilians at a rate sufficient to keep three brigades in individual training simultaneously, accessions will become the bottleneck in an expansion system with four operating NTCs.

If recruiting (or conscription) is not a problem, then the bottleneck is the industrial base. If the industrial base were to increase to maximum production levels (two plants and 2.5 shifts) for each of the main heavy equipment elements, it could stay up with the training output of four NTCs (after the second plant has come on line) but would lag behind the training output of five or more NTCs.[5]

---

[5] If it is operating at nominal capacity with one plant and 2.5 shifts from a hot base, it will be the bottleneck shortly after the complement of initial equipment runs out if there are two or more combat training centers.

SUMMARY

There is currently only one CONUS-based advanced or brigade-level training facility in operation (at the National Training Center). If that were to remain the only combat training center, it would be the primary bottleneck to expansion and little that could change in the future would alter that fact. The expansion rate would be similar to that in Figure 4.2. Reducing the advanced training time would increase the overall expansion rate. On the other hand, it is entirely possible that the advanced training time could increase in the future because of greater coordination requirements (e.g., due to internetted communications), more sophisticated equipment (requiring longer training periods), a greater variety of equipment (requiring more coordination), etc. Anything that increased the advanced training time would depress the expansion rate in Figure 4.2.

Adding another NTC doesn't entirely remove training as the bottleneck, though there are several scenarios in which the industrial base can become the bottleneck. This includes scenarios in which today's equipment "surplus" has dissipated through obsolescence, attrition, or other neglect. In addition, with two NTCs it is easier for individual training to become the bottleneck.

For today's dominant parameters (initial equipment stockpiles, nominal production rates possible for that equipment, and the civilian population with recent military experience), three NTCs would produce a generally optimal expansion system in the sense that any dropoff in other parameters would produce a dropoff in the overall output of the system. In some cases, that dropoff doesn't occur until four years into the expansion; in other cases the dropoff is immediate.

Building four or more NTCs could produce some benefit in some situations, but it generally produces diminishing returns. If today's individual training system can be maintained at full capacity, the bottleneck with four or more NTCs would be the industrial base.

Speaking more specifically about the industrial base, the nominal full-up situation of one plant, 2.5 shifts producing each of the major heavy division equipment elements can keep up with the training at two NTCs. It would take two plants and 2.5 shifts (once they were on line) to stay up with or ahead of the training at three or four NTCs.

Beyond that, our correspondents thought it unlikely we would ever go to three plants, meaning the industrial base would quickly become the bottleneck in any scenario with five or more NTCs.

If the current stock of equipment fails to complete its predicted life span, the industrial base must be brought into the system sooner. If all of one type of equipment were rendered obsolete by a surprise breakthrough, that element of the heavy forces would immediately be affected by the industrial base as the Army scrambled to produce a more robust alternative. Other than that, one would expect the current equipment, perhaps with modifications and perhaps with improved versions downstream, to last out its projected life. If a fraction of it were somehow to be taken out of the inventory, the effect of that fraction could be described directly for a given scenario, but it wouldn't appear useful to characterize such an occurrence in general.

Besides NTCs and the industrial base, there are two elements of the expansion system that can cause problems for virtually any expansion scenario. One is well known and situation dependent, the other is largely unknowable and probably situation dependent. The first is retraining time. If brigades produced by the expansion are not sent into the field, their readiness will slowly erode to a point where they will need retraining. If that is the case for a given expansion, the point at which the first brigade will need retraining becomes the point at which the expansion system will become saturated with retraining needs (unless new NTCs are produced).

The second element that can cause problems is the ability to attract new soldiers. Expanding beyond today's 57 ready and reserve brigades (or a different total in the future) will require new volunteers or draftees. All the explorations beyond 57 brigades presumed an ability to keep the current individual training facilities at full capacity training new recruits in individual soldiering skills. It is one thing to prepare the facilities for individual training and quite another, under today's conditions, to be able to attract the volunteers to keep those facilities busy. And the Army's ability to attract those volunteers could depend strongly on the reasons for the intended expansion. There were several scenarios where the inability to keep the individual training facilities full caused recruiting to become the bottleneck in the expansion system. Without assuming the timely reinstitution of a draft, the process of recruiting volunteers must be considered a potential (and difficult-to-predict) bottleneck to any expansion that goes beyond the ready and reserve forces in place.

Chapter Five

COSTS AND EXPANDING THE ARMY'S HEAVY FORCES IN THE FUTURE

INTRODUCTION

During peacetime, the dominant concern about expandability capabilities is their recurring costs; during an actual expansion, the primary concern is the time it will take to expand to the appropriate size. Given the unlikelihood of the Army's having to expand in the near future, it would be natural to keep the recurring costs of an expansion capability low. History would look harshly, however, on an Army that kept the capabilities so low that it was unable to expand sufficiently and quickly enough during an actual crisis. The primary tradeoff of strategic planning interest, then, places the recurring costs of an expansion capability during peacetime against the speed with which the Army could expand in the future starting with those capabilities.

This is not a straightforward costing issue, then, and a more classical costing analysis is thus not appropriate for our purposes. We chose instead a very narrow, specifically targeted costing analysis. Trading costs during peacetime with speed of expansion (at an unspecified time for unspecified reasons) is an analytical challenge. The fundamental question can be made more general, however, in a way that permits an exploratory analysis to provide insights. In particular, in this chapter we try to answer two questions:

- What can be said (both now and in the future) about the Army's ability to expand to a given force size in a given time that minimizes recurring peacetime costs?

- What can be said (both now and in the future) about the Army's ability to expand to a given force size in a given time that pays attention both to recurring peacetime costs and expansion costs?

To explore these questions, we will be interested in a different aspect of expansion. In the timing analyses we were interested in the timelines associated with bringing additional capabilities on line. In the costing analysis we will be interested in using each point in a space defined by number of brigades and time as the endpoint of an expansion. That is, for a given number of brigades, say 40, and a given time, say 12 months, we will be interested in the expansion capabilities and peacetime structure that would be required to be able to expand to 40 brigades within 12 months. We will call each such size and time target a "point in expansion size-time (state) space." This size-time space represents a large set of plausible expansion sizes and expansion times. The questions we ask will be about how the recurring and expansion costs vary over that size-time space.

The first question above does not deal with costs during an expansion. It addresses the minimum recurring costs of today's forces that would permit an expansion (with such additional capabilities deemed necessary to be brought on line during the expansion) to a given force size in a given time (or point in expansion size-time space). By being able to look (through exploratory modeling) at many of the points in expansion size-time space and being able to see how that landscape changes as parameter values are varied, we will be able to characterize any peculiarities in the cost of expandability. For example, if there are regions in size-time space in which the recurring costs required to reach them either drop or rise dramatically, these would be important regions to understand further—what is it about the target size or target time that causes things to change so quickly? Conversely, if the recurring costs required to reach any given size in any given time don't vary much or vary "smoothly" over the entire space and over the parameter ranges, it suggests that one's instincts about expansion costs are probably pretty good, and that there are likely to be few surprises with respect to expandability in the future.

It would be defensible to ignore expansion costs during a large expansion such as occurred during World War II. That is, it has historically been the case that large expansions have signaled a security threat so severe that the nation was willing to incur the costs of the expansion. For smaller expansions, the costs of the expansion itself are more likely to be a concern. This is what the second question addresses. In this case, we were again working in expansion size-time space and asking how we might achieve a given target size in a given target time that pays attention to both the recurring costs during peacetime and the costs during the expansion. In this case, the exploratory model varied Army designs until it found one that met the target size and time and minimized the recurring costs during peacetime. It saves the force structure and the capabilities brought on line during the expansion, permitting an examination of the associated expansion costs and other characteristics of the minimal-recurring-cost force structure.

THE COST OF EXPANDING HEAVY FORCES EQUIPMENT

The industrial base is likely to be maintained in its current "warm" condition with one plant running one shift to upgrade units for all the major weapon systems. The recurring costs of the industrial base, then, are likely to be relatively constant (in fixed-year dollars).

For expansions to 78 brigades or less, the industrial base would not be required to build new units (though it would undoubtedly be pressed into service for any sizable expansion). If more than the currently available 78 brigades of equipment were required for an expansion, the industrial base would have to be brought to a "hot" condition. There would be about a two-year delay to build new units because certain critical suppliers are no longer available, such as the supplier for the Abrams engine. In this case, the costs of building new units would be considered an expansion cost. Expansion costs would then be dominated by the cost of providing new equipment for new heavy units. Some measure of this is evident in the following table:[1]

---

[1] These figures come from the latest version of the Force and Organization Cost Estimating System (FORCES) model provided by the Army's Cost and Economic Analysis Center (CEAC). FORCES is a cost model that incorporates detailed database information on all units in the current force structure, both active and reserve. The FORCES model is capable of estimating the cost of different events in a unit's life cycle, including acquisition of resource costs and annual operating costs. The fixed cost of a division was estimated using the FORCES Acquisition of Resources cost

Unit Type	Representative Unit	Total Fixed Cost	Equipment Acquisition Cost	Personnel Acquisition Cost	Recurring Cost
Heavy active	1st Armored Division	$1,447	$1,410	$37	$405
Heavy ARNG	35th Mechanized Division	$912	$881	$31	$82

All costs are in millions (FY98) for a brigade.

The fixed costs of acquiring equipment for an active heavy division are roughly 40 times the acquisition costs of recruiting and training the personnel to man the equipment. This disparity washes out any interesting results when industrial base costs are included in a costing analysis.

If, as during the Reagan buildup, the industrial base is set to the task of producing new equipment, this recapitalization of the Army could be accurately labeled neither as an expansion cost, if it happened to overlap with an expansion, nor as a recurring cost that could be ascribed to an expandability capability.

In sum, in looking at the recurring costs of an expandability capability and the costs of an expansion, the industrial base plays as either a fixed cost in the recurring-cost equation, a dominating cost in the expansion-cost equation, or an "orthogonal" cost if a modernization of equipment happens to overlap with an expansion. In all cases there is little to gain from having included the industrial base in this type of narrowly focused costing analysis. For this reason, the industrial base will not be included in the costing analysis.

The recurring costs for a "warm" industrial base are real costs, but they will be the same across all the costing scenarios. Since we are

---

estimate. Acquisition of Resources is defined as the cost to procure the material and personnel required for the selected unit. It includes the costs of recruiting and training personnel through initial MOS training. The recurring cost of a division was computed as the FORCES Annual Operations cost estimate. Annual Operating costs is defined as the direct and indirect costs to operate the selected force at the specified readiness and optempo levels. For a detailed description and breakdown of these cost estimates, refer to *The FORCES Cost Model (FORCES 97.1), Introduction*, June 1997, Department of the Army, CEAC.

interested in relative costs, even these recurring costs can and will be ignored in what follows.

While justified on analytic grounds, this also relieves some of the computational burden. If fixed and recurring costs were added to the model for each of the timing parameters, the model could triple in size. This would seriously slow the computations and adversely affect the interactive nature of the explorations. As described below, one part of the costing analysis will deliberately slow the computations in order to do optimizing searches as it is. Eliminating the industrial base parameters from the model ensures adequate interactive speed for some of the costing explorations and enables adequate speed in the optimizing searches of the remaining explorations.

DEFINITIONS

Before proceeding, it is important to understand the definitions we are using for recurring costs and expansion costs.

Recurring Costs

What are the recurring costs associated with expandability? In a narrow view, they are the costs of those units and facilities that would be used in an expansion. For an expansion to a given size (around which all our costing explorations will be centered), this includes the recurring costs for those units that become part of the expanded force. For example, if a given expansion requires all the ready units and half the reserve units, the recurring costs will include only those for half the reserve units because the remainder weren't needed. While this very narrow view would require prescience if demanded in practice, it provides a consistent set of recurring costs when one is looking across the entire expansion size-time space.

Another recurring cost is for any training facilities (such as the NTC at Fort Irwin) being maintained before an expansion begins. If, during the subsequent expansion, enough additional NTCs are opened to require ready units in order to supply sufficient trainers, those must have been ready units at the beginning of the expansion. That is, the Army has been adamant that NTC trainers come only from the active, ready Army (i.e., that trained-up reserves are inappropriate for this mission). This requires that there have been sufficient ready troops at the beginning of the expansion to man the additional NTCs, and the costs of those ready units must also be counted in the recurring costs.

It is important to note that in a full analysis that included the industrial base, we would also want to consider the recurring costs of a warm industrial base and the recurring costs of equipment maintenance. Because we have already dismissed the industrial base and because our aim is relative costs, these recurring costs can be ignored without affecting the results of our comparisons.

Expansion Costs

For our purposes, expansion costs will generally be those costs in our simplified model that would not ordinarily have been incurred if expansion had not occurred. This includes the fixed and recurring costs of any training facilities that are added after expansion began as well as the cost of the trainers for those facilities. It also includes the added costs incurred by having ready troops instead of troops in reserve. That is, had a unit stayed in reserve, it would have remained at its recurring cost level. Once it becomes trained and ready, however, it takes on the costs of a ready unit. The difference in those two costs (over the time it is maintaining its readiness before the target size has been reached) is an expansion cost.

THE COST MODEL AND PARAMETERS

As above, we are ignoring the costs of equipment in this exercise. In addition, because of the way we defined recurring and expansion costs, we ignored a great many other costs. This left very few costs to consider and vary. We distinguished only between fixed and recurring costs. Fixed costs were interpreted as the cost required to produce a capability initially. These are roughly comparable to capital costs (or plant replacement value). Recurring costs are those required to maintain a given capability for a year. These are comparable to annual operating and maintenance costs. The costing numbers used for the heavy brigades costing model can be found in Appendix B.

RECURRING COSTS VERSUS THE TIME TO EXPAND HEAVY FORCES

For the costing explorations, we will be interested in points in expansion size-time space—that is, we will be interested in the recurring costs of expanding where there is a target size for the expansion and a target time allowed in which to complete the expansion. From earlier, it is trivially possible to reach a given expansion size goal by having that many ready troops before the call for expansion comes. In computing the recurring costs, the exploratory model allows for any of the model parameters to be set beforehand except for the initial number of ready brigades. It then determines the number of initial ready brigades required to reach the target size by the target time, returns that as an output, and computes the recurring costs that include the required ready brigades. That is, the model takes the position that, in the best of worlds, what is paramount in an expansion is reaching the target size in the target time, and it computes what is required to do the job.

For the most part, the exploration of recurring costs was uninteresting in the sense that there were few surprises. Changing a given cost parameter generally changed the overall costs in a "smooth" and obvious way. Changing two or more parameters at a time had much the same effect.

There was one set of parameter values that produced a "kink" in the graphs of recurring costs. An example is shown in Figure 5.1. The story that this graph tells is this: *If the recurring cost of a ready brigade is significantly larger than that of an ERB, a NG brigade, or a brigade of civilians*, one can find parameter values for which there are sharp breaks in the recurring cost curves.

Observations. Figure 5.1 shows the recurring costs for reaching targets of 15, 30, 45, and 60 brigade targets. The model parameter values are all at nominal levels (except for the maximum number of NTCs that could be built during the expansion).[2] There is a clear "knee" in these cost curves.

_____

[2]The nominal value is set to six but was restricted to three for this case.

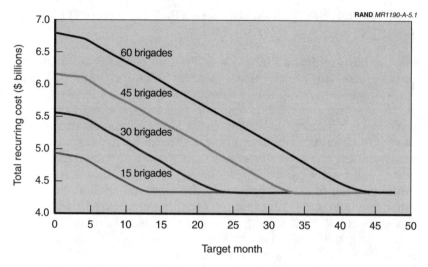

**Figure 5.1—Recurring Costs as a Function of the Number of
Target Brigades and Target Month**

The nominal recurring cost of a ready brigade is about four times
that of an ERB and about five times that of a NG brigade. Lowering
that ratio "flattens" the curves in Figure 5.1. Even at equal costs for
ready, ERB, and NG brigades, a slight knee exists when civilian
brigades are brought to readiness.

Figure 5.2 shows the effects of changing the maximum number of
NTCs allowable during expansion. The peak at the beginning of the
curves for the higher numbers of NTCs represents the ready brigades
required to man the additional NTCs. The slight wiggle at the end of
the curve for six NTCs is indicative that the nominal individual
training site capacity is starting to slow down the expansion rate. If
the number of individual training sites is lowered, or the advanced
training time is lowered, the wiggles in the curves become more pronounced and the "knee" abates.

The remaining variables do not measurably affect the cost curves.
For example, changing the starting reserve structure (ERBs and NG
brigades) through the entire parameter range changes the total
recurring costs but does not change the shape of these curves.

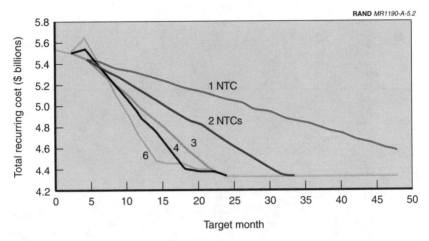

Figure 5.2—Recurring Costs as a Function of the Number of NTCs and Target Month (For a 30 brigade target size)

Interpretation. The longer the Army has to expand, the less expensively (in recurring costs) it can be done because it does not need as many (expensive) ready troops to meet the target. At some point it doesn't need any ready troops to meet the target force in the target time. This can be seen clearly in Figure 5.3, which shows the number of ready troops required to meet the goals in Figure 5.1. Recurring costs drop as fewer ready troops are required to meet the target size in the target time. At the point that no ready troops are required to meet the goal, the cost savings with more time disappear because the system is meeting its target goals with civilian brigades. The larger the target size, the longer the target time must be before no ready brigades are required and the recurring costs level off. As the relative recurring costs between ready and reserve diminish, so do the savings and the knee in the cost curve.

The changes become more precipitous with more NTCs because the increased training capacity means the number of ready troops required to meet size targets drops quickly with time. With fewer NTCs it takes a long time to train up reserves, requiring ready troops to meet size targets for a longer period. Eventually, these curves would show a slight "knee" as the costs leveled off.

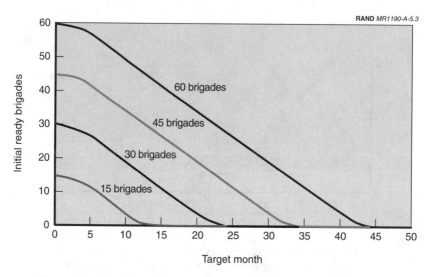

Figure 5.3—Number of Initial Ready Brigades as a Function of the Target Brigades and Target Month

If there are fewer reserve troops and the individual training capacities are at or less than nominal, meeting size targets will require more ready troops for longer periods, pushing the knee out farther in time and lessening its sharpness. There will still be a slight knee in these curves because civilian brigades have "no" recurring costs,[3] so if all ready troops could be created from civilian brigades, the recurring costs would be less than those for readying reserve brigades.

These findings are generally unsurprising and rest on the disparity in recurring costs between ready brigades and reserve brigades. Further, there is an "accountant's assumption" that reserve or civilian brigades brought to readiness are the equivalent of ready brigades. From a military perspective this is a dangerous assumption, and it is exacerbated when considering timelines that are long enough (or expansion systems that can ready troops quickly enough) to allow most of the ready troops to have been generated from

---

[3]In reality, there are costs associated with being able to recruit civilians or to conscript them, but these are significantly less than the costs of maintaining a reserve brigade.

reserves or civilians. On the other hand, for timelines so short that reserves cannot be readied in time, it makes the militarily comfortable point that the only way to meet the challenge is through maintaining sufficient ready brigades.

RECURRING AND EXPANSION COSTS

In peacetime, the recurring costs of maintaining an expansion capability are more important than the expansion costs the Army would have to incur if an expansion became necessary. Because of this, in comparing the recurring costs of a given system with the expansion costs of that system, one would want to weight the recurring costs in some way. Rather than pursue an appropriate weighting scheme, we took a different approach. For a given point in expansion size-time space, we asked, "What overall force structure and expansion system meets the target goals with *minimum* recurring cost?" Having that force structure we could then ask what the expansion costs would be for that system and whether there were any particularly outrageous expansion costs that would be incurred if one were to attempt to minimize the recurring costs of the expansion system.

In explaining the results of this approach, we will again look at them by the number of NTCs, but with a slight twist. In this case, the number of NTCs is the maximum number allowable in an expansion. That is, the exploratory model looked for the optimal force structure and expansion system that minimized costs and may not have chosen to build all the NTCs allowable. It was, however, constrained as to the total number it *could* build.

Three NTCs

Observations. Figure 5.4 shows an approximation of the minimal recurring costs in size-time space. That is, there is an infinitude of points in size-time space, so it would be impossible to calculate the minimal-recurring-cost solution for each point. We divided the space into a 50×50 grid and calculated the minimal-recurring-cost solution for the centroid of each box in the grid—giving the approximation shown in the figure.

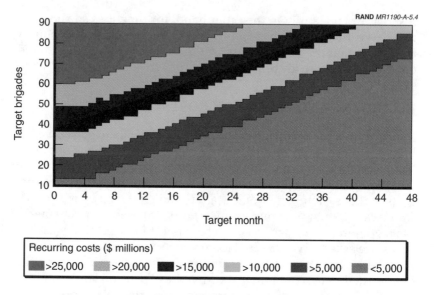

**Figure 5.4—Minimal Recurring Costs with No More Than
Three Allowable NTCs**

Again, these are not the full recurring costs of the expansion capability. It is not the costs per se that we are interested in, but the shapes of the regions and how they change with changing circumstances.[4] These graphs are, however, a good representation of the change in recurring costs as the target size and allowable time change. As would be expected, if a large Army is required quickly, the recurring costs rise rather steeply as the target size increases.

The recurring costs are driven primarily by the number of ready brigades required to meet the expansion target on time, as shown in Figure 5.5. Not surprisingly, the slope of the cost lines is reminiscent of the slope of the line for readying reserve troops with three NTCs (reproduced here as Figure 5.6).

_____

[4]These plots are much more time-consuming than the timing plots. Each requires 2,500 searches for minimal-cost solutions. Time and resources did not permit a full study of how they change with changes in other variables. In some cases it will be clear how other parameter values would affect the results, and these will be described.

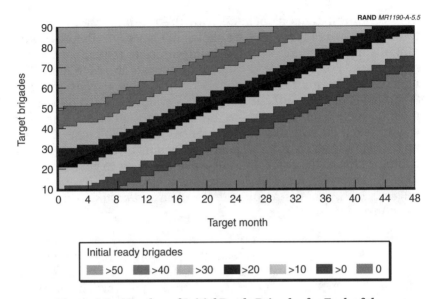

Figure 5.5—Number of Initial Ready Brigades for Each of the Minimal Recurring Cost Force Structures (with no more than three allowable NTCs)

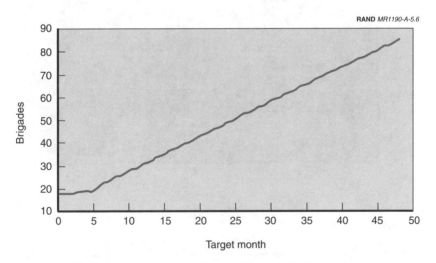

Figure 5.6—Expanding Heavy Forces with Three NTCs

The expansion costs associated with the minimal-recurring-cost force structures of Figure 5.4 are shown in Figure 5.7. Before beginning the analysis of Figure 5.7, it is important to remember what it represents. For example, it suggests the costs to reach 80 target brigades in four months is less than $5 billion, while the cost to reach 80 brigades in 12 months is between $5 billion and $10 billion. It must be remembered that these are the *expansion* costs. The only way to reach 80 brigades in four months is to have most of those brigades in the active force. There is a huge recurring cost of maintaining a large active force, but the expansion costs to get that active force to an active force of 80 brigades are modest. To reach 80 brigades in 12 months requires fewer recurring costs but more expansion costs.

There are two major themes in this figure. The first centers around the vertical stripes. They represent the fact that trained reserve and civilian brigades are carried as an expansion cost. As a function of time (and independent of other factors), reserve and civilian brigades can be readied and begin adding to expansion costs. The vertical stripes represent regions of these costs.

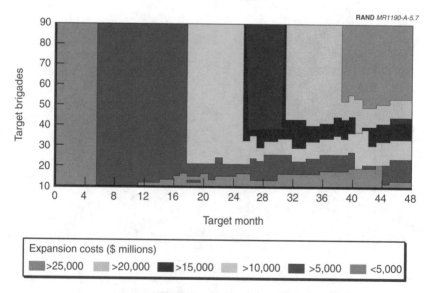

Figure 5.7—Expansion Costs for Each of the Minimal Recurring Cost Force Structures (with no more than three allowable NTCs)

The second theme centers around the slowly growing regions of reduced costs as time for expansion gets large enough. Heretofore, the serrations in the edges between colored regions have been due primarily to having divided the expansion size-time region into a 50×50 grid. In Figure 5.7, some of that serration represents the interaction between the increase in expansion costs due to increased numbers of readied reserve and civilian brigades and the reduced costs from needing fewer NTCs in order to meet the size goal. Figure 5.8 shows the initial number of reserve brigades in each of the minimal-recurring-cost force structures. The banding evident in Figure 5.7 is evident here, as is the slowly reduced need even for reserve units as less expensive civilian units can be readied.[5]

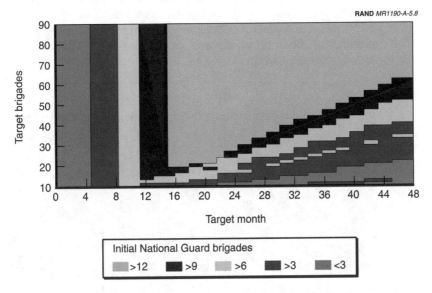

Figure 5.8—Initial Reserve Brigades for Each of the Minimal Recurring Cost Force Structures (with no more than three allowable NTCs)

---

[5]The colored "islands" in some of the regions of Figure 5.8 are due to interactions between discretization of the expansion process, the 50×50 discretization of the size-time space, and the optimization routine that computed the minimum-cost solutions. While adding interest to the picture, they do not materially affect the interpretation or conclusions.

Figure 5.9 shows the number of NTCs in each of the optimal force structures. This shows the decreasing need for additional NTCs in order to meet the size-time goals. The uneven conjunction of this with the reduction in the required number of reserve brigades leads to the interesting patterns in Figure 5.7.

Interpretation. Over most of the expansion size-time space, recurring costs and expansion costs vary smoothly. That is, even for optimal recurring costs, there are no sharp changes in the resulting expansion costs.

There is a bit more variability in the recurring costs than in the expansion costs. Recurring costs vary by a factor of five in the heart of the size-time space, while expansion costs vary by only a factor of three. Recurring costs are highest when large forces are required in a short period, while expansion costs generally increase with the time of the expansion. Both recurring and expansion costs can be kept low for small, slow expansions.

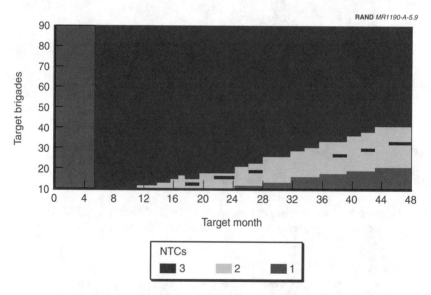

Figure 5.9—Number of NTCs for Each of the Minimal Recurring Cost
Force Structures (with no more than three allowable NTCs)

Six NTCs

Observations. Figure 5.10 shows the recurring costs of an expansion system that can build up to six NTCs. In comparison with Figure 5.4, the bands for the six-NTC case rise more steeply as a function of the target size (and thus they drop off somewhat more precipitously as a function of time).

Figure 5.11 shows the number of ready brigades required to meet expansion targets on time for the six-NTC case. Again, it drops off more quickly as a function of time.

Figure 5.12 shows the expansion costs associated with the minimal-recurring-cost force structures of Figure 5.10. While showing much the same character as the expansion costs in Figure 5.7, the basic difference here is how quickly the bands rise as the five new NTCs come on line and new brigades are readied at a faster rate. Figure 5.13 shows the number of NTCs used by each of the optimal force structures of Figure 5.7. It is identical to Figure 5.9 for short-time ex-

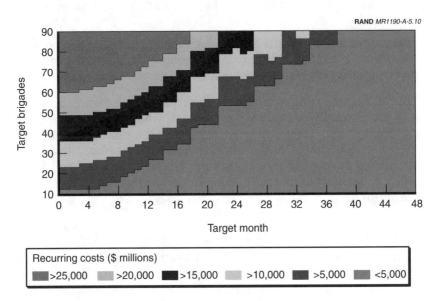

Figure 5.10—Minimal Recurring Costs with No More Than Six Allowable NTCs

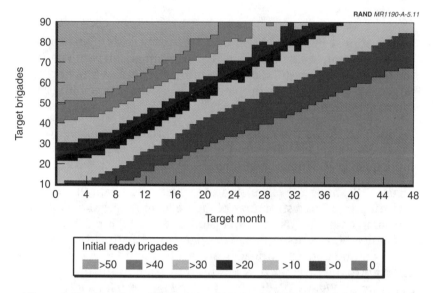

Figure 5.11—Number of Initial Ready Brigades for Each of the Minimal Recurring Cost Force Structures (with no more than six allowable NTCs)

pansions or small expansions. Where the expansions are larger and more time is available, bringing on more NTCs allows for smaller recurring costs while still meeting the time target.

The rather interesting striping and island patterns in Figures 5.12 and 5.13 are primarily caused by the timing of when brigades enter advanced training. If they all enter at roughly the same time, as here, there can be "jumps" in the number of ready brigades, causing an optimizing program to take advantage of those cases when one fewer NTC can be used to meet the target force size by being able to time the jumps right.

These could easily have been smoothed out by changing the timing on entering brigades, but they were left in to show that somewhat surprising things can happen even in a simple pipeline model such as this when one can get even a crude picture of the entire expansion size-time space. These are real effects in the sense that if the sequencing of training is just right, one could optimally meet an expansion goal with six NTCs, while a slightly larger goal could be met in less time with the same sequencing and only five NTCs. They

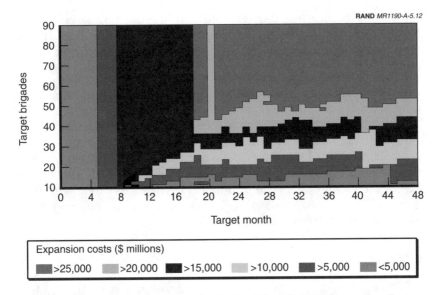

**Figure 5.12—Expansion Costs for Each of the Minimal Recurring Cost
Force Structures (with no more than six allowable NTCs)**

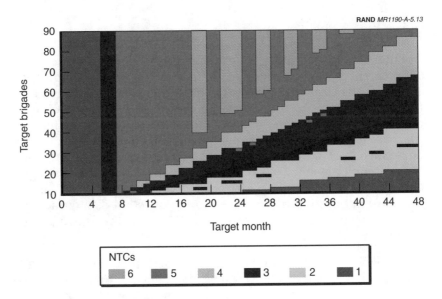

**Figure 5.13—Number of NTCs for Each of the Minimal Recurring Cost
Force Structures (with no more than six allowable NTCs)**

are not real in the sense that it would be difficult to take advantage of this fact in planning because of the great uncertainties both in the target size of an expansion and in the available time in which to complete it.

Interpretation. Recurring and expansion costs for short-time contingencies are unaffected by the number of NTCs that might be built during an expansion. There are large regions in the long, slow expansion area where the maximal number of NTCs that can help an expansion is low. This obviates the need for a large number of additional NTCs during an expansion for those cases. Adding several NTCs, then, can reduce the minimal-recurring-cost force structure required to meet expansion goals, but only for expansions whose goals are long enough and large enough.

CONCLUSIONS

This has been a very rudimentary cost analysis aimed at getting a feel for the tradeoff between the recurring peacetime costs of an expansion system and how they affect the speed of an expansion—should it occur.

There are few surprises in the costing arena.

The industrial base is likely to be maintained in a "warm" condition, with current plants being used to modernize the current equipment. The recurring costs of the industrial base, then, are likely to remain relatively constant. If more than the current 78 brigades of equipment are required for an expansion, the costs of turning the industrial base to production of new equipment will dominate the costs of an expansion. If a warm industrial base is set to the task of producing new equipment, it would likely be coincidence that made it happen around the time of an expansion. The industrial base costs, then, are not particularly germane to the question of the tradeoff between recurring costs during peacetime and the speed of an expansion.

It is unlikely that more than one heavy-force combat training center (the National Training Center) will be maintained during peacetime. The primary means for replacing the need for—and expense of—high-cost, ready brigades is the ability to train up reserve brigades.

For periods of less than about half a year, the expansion system cannot add much in the way of readied brigades. Similarly, for expansions smaller than six or so brigades per year, the current ability to expand is adequate and only reserve troops are required to meet timelines. It is in between these extremes that an expansion capability beyond today's normal training capability can add cost-effective expandability.

As long as the recurring costs of a ready brigade are much greater than those of a reserve or civilian brigade, the time and ability to ready a reserve or civilian brigade would permit a substantial reduction in the recurring cost of providing the expansion capability.

There is in this, however, an implicit assumption that reserve or civilian brigades brought to readiness are the equivalent of ready brigades. Further, it suggests that for some expansions, the Army could manage with just enough ready brigades to man the training sites and could train reserve (or even civilian) brigades in time to meet the target. This is the logical extreme of taking a pure costing approach to the "optimal" size and force structure of the Army. From a costing standpoint, the minimum Army could be as small as two or three brigades composed of trainers. No serious military analyst would accept that as a reasonable answer, but it provides a logical minimum for an Army that could be expanded from a trained base.

Heavy forces are readied in large discrete units. Because of this, the exact time that brigades come out of advanced training at several training sites can make a difference in trying to compute the optimal number of training sites to bring on line. While this is a real effect, it would be fruitless to try to take advantage of it because of the serious uncertainties in the parameters affecting that timing. It does, however, suggest that—even if the size and timing of an expansion could be known—computing an "optimal" expansion system for that expansion would make little sense.

WORRY CURVES

Perhaps the most interesting use of the graphs generated in the cost analysis is to use the plots of required initial ready brigades to overlay a drawing of one's "worry curve." This is a curve that describes the time-phased requirements that one worries the Army might need now or in the future. For example, if one worries that we might need as many as 15 heavy brigades in a very short time, that represents several points on one's worry curve. What forces could conceivably be needed in a one-year period? Two years? Four? In this way, one can draw one's worry curve on a plot of the initial ready brigades required for one's preferred set of expansion parameters. Figure 5.14 shows one example of such a worry curve overlaid on Figure 5.5. In this example, the worry is that while we wouldn't need more than 10 divisions immediately, we might have to build up to as large as 19 divisions in a year, but not more than 23 divisions in four years. The graph indicates that with an expansion system that has three NTCs, the Army would have to maintain just over 40 ready brigades in order to meet all the worries on the worry curve.

It's a visual way to both gauge the number of ready brigades required for any expansion contingency and a way of testing that requirement over a variety of plausible changes in the future. This is a more personal, rather than analytic, use (unless one can justify analytically a specific worry curve), but it provides great insight into both required current force levels and expandability considerations.

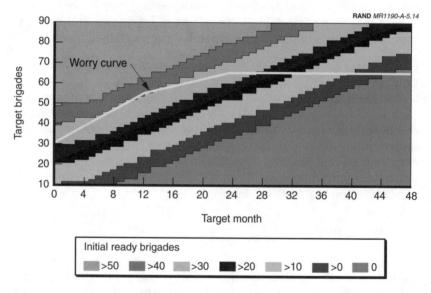

Figure 5.14—A Worry Curve

Chapter Six

EXPANDING THE ARMY'S LIGHT DIVISIONS TODAY

A major expansion of U.S. forces has usually been associated with a threat to vital U.S. interests from a peer competitor. That type of expansion usually means expanding heavy forces. However, the days of meeting an adversary on an open battlefield such as in World War II or in the Persian Gulf War may be over. At least for the foreseeable future, such an encounter is unlikely. That is not to say that such an encounter could not happen. As we have argued above, the United States should be prepared for such an encounter sometime in the future and thus must prepare now. But a more likely requirement for expansion of U.S. forces in the shorter term is the expansion of light forces to conduct a different type of mission.[1]

The end of the Cold War ended a major confrontation with a peer competitor but brought to the fore many other conflicts. Whereas the Soviet Union threatened vital U.S. interests, regions of potential instability have grown in which important but perhaps not vital U.S. interests are threatened, such as Southeast Asia, South Asia, North Africa, the Middle East, Eastern Europe, and the Caspian Basin. Closer to home, there are areas of instability throughout Central and South America.

---

[1]The trend in the use of armored forces since World War II is clear. Worldwide, the only major armored conflicts since 1945 occurred between Israel and its neighbors and between India and Pakistan. The Korean War and the Vietnam War saw the widespread use of light infantry forces and little use of heavy ground forces. The Chinese and Indians fought with infantry in 1962, and even the Iran-Iraq War in the 1980s was basically an infantry war.

Instability in some of these regions would be sufficiently important that the United States might respond to try to restore stability and order. Restoring stability and order to an area of chaos could well happen in densely populated areas and involve combat under restricted rules of engagement. It is these types of missions that we will consider in this chapter on light force expandability.[2]

We first analyze the light force expansion process. We examine the potential expandability of advanced training for light forces and determine the availability of personnel to undergo that training. Finally, we compare the potential training capacity with unit availability under nominal assumptions.

EXPANSION PROCESS

As with the heavy forces, we chose to examine the process up to a nominal 40 light divisions. Expanding to this level should identify all important issues that arise during the expansion process. Furthermore, expanding to 40 divisions will enable us to compare the expansion rate of light forces with the expansion rate for heavy forces estimated above. Beyond a certain force level, the expandability process itself will be the same.

We explore how long it would take to expand to 40 light divisions today and describe the processes involved. The general light force expansion process can be characterized by the flow model outlined in Figure 6.1. Although the actual process has been greatly simplified in the model, the major elements involved in training and equipping the force are represented.

This flow model of the expandability of light forces is similar to the flow model for heavy force expansion. Two of the elements are the same: (1) the training process, facilities, and training personnel, and (2) the individuals or units that are transformed into ready units by the training process. However, the nature of the training process and

---

[2]For a discussion of this new challenge, see, for example, Ralph Peters, "Our soldiers, Their Cities," *Parameters*, Spring 1996, pp. 43–50.

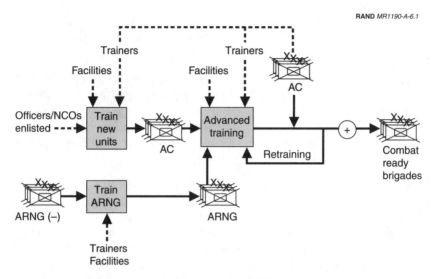

Figure 6.1—Light Force Expandability Flow Model

the units to be trained will be different. Furthermore, equipment production is omitted from the flow model for light force expandability because it is not likely to constrain light force expansion.

The first element in the flow model—the training processes—is represented in Figure 6.1 by the three training boxes. To train new units, enlisted personnel receive basic and MOS training, and NCOs and officers are trained or brought in from active units to lead them. When this training is completed, AC units are formed that are ready for advanced training. ARNG units would have to fill individual and unit deficiencies before they were ready for more advanced training. Advanced training at the battalion and brigade level is the final training for all units and must be completed before units are combat ready.

The process of training requires training personnel, support personnel, and facilities. A sufficient number of all types of training personnel must be available for effective training. In the model, the AC forces supply these trainers if they are otherwise unavailable.[3]   In addition, training facilities of different types must be available, which depend on the type of training required.

The second element in the flow model is the individuals or units that receive the training. There are three types of forces that may receive training: active component forces, Army National Guard brigades, and newly trained brigades. As shown in Figure 6.1, each of these individuals or units enters the expansion process in a different place, depending on its readiness for training.

We have omitted equipment expansion from this flow model. Compared to outfitting a heavy division, the light division equipment requirement is relatively modest. There are three main equipment items besides small arms that a light division includes in its TOE: trucks, HMMWVs, and artillery, either howitzers or MLRS. From the company command up to the brigade command, the heaviest equipment item is the five-ton truck. DIVARTY usually has towed or self-propelled 105 and 155 howitzers and possibly MLRS batteries assigned. MLRS is also a corps-level asset that can complement a division's organic artillery. Producing sufficient MLRS could prove to be a problem but we have addressed this production issue in Chapter Three.

The first two major subsections below take up these two main elements of the expansion process—the training process and the trainees. These sections will address both the status of the system today as well as its ability to grow if a large expansion was required. The third subsection estimates the maximum ability of the expansion system to produce trained light forces.

TRAINING PROCESSES, FACILITIES, AND PERSONNEL

The three elements of the training process required for expanding are advanced training (which all units must undergo), and the training for both ARNG and new units before they are ready for this advanced training. We consider each in turn.

---

[3]We assume that IRR and ARNG trainers may be needed as fillers to mobilized ARNG divisions and for ARNG training.

Advanced Training

In the expansion model depicted in Figure 6.1, we assume that all units must undergo advanced training at a combat training center (CTC) before they are ready for combat. For light forces, this CTC will be the Joint Readiness Training Center (JRTC) or a similar training facility. Currently, in peacetime, individual soldiers join existing units after receiving basic and MOS training. In the event of an expansion before an imminent conflict, newly accessed personnel could not blend in with established units to receive on-the-job training (OJT). Rather, they would have to be trained as units and made ready for combat. To prepare for combat, these units would require final brigade-level training.

Activities and timing of advanced training. The mission of the JRTC or another similar facility is to train Army light infantry and special operations units to conduct light infantry operations, including training for urban combat. The goal is to create realistic joint and combined arms training. The JRTC meets this goal through the use of MILES-supported field training exercises (FTXs) that include force-on-force exercises and live fire exercises. The JRTC tailors all exercises to accomplish the specific training goals established in the unit commander's Mission Essential Tasks List (METL). Although specific operations in a rotation may vary due to a commander's METL, the standard tactical operations include forced entry, movement to contact, defense, deliberate attack, MOUT, and air assault.

Planning for the rotation by the brigade commander and his staff begins about six months before the actual rotation begins. At this time, the brigade commander and his staff determine unit training objectives and begin planning the rotation. The plan is made and refined before the rotation begins.

Typically, there are three phases to the actual rotation: the first phase is usually some type of low-intensity operation, the second phase a defensive operation, and the third phase an offensive operation. Specific missions for these phases are tailored to accomplish the training goals established in the unit commander's METL.

The low-intensity conflict portion of the rotation usually lasts for three or four days, during which initial reconnaissance and insertion of SOF occurs. Then the regular light force (two battalions plus usually a CPX for the third battalion) arrives to fight the insurgency for several days. The next portion begins with a larger-scale invasion in which the battalions engage in more conventional warfare, first a defensive phase and then an offensive phase. Usually included in this conventional warfare phase is a MOUT engagement.

A complete rotation usually consists of 13 days "in the box" and four to five days of planning, for a total of about three weeks per rotation. This process could be accelerated depending on what activities are to be emphasized. For example, the original low-intensity conflict phase could be eliminated and emphasis placed on the more conventional offensive and defensive aspects of the unit training. We have not examined these activities to determine if some could be run in parallel to increase their rate even further.

Advanced training inputs. This advanced training process must have adequate facilities and a sufficient number of capable trainers to effectively conduct advanced unit training. First we describe the facilities.

- **Requirements.** Advanced training has a number of requirements for the facilities where such training is to occur if the training is to be effective. Because a future conflict is likely to include operations in an urban environment, a realistic MOUT complex should be available with training for all soldiers. In addition, all training should take place at a site with adequate infrastructure, which includes communications, laser engagement systems, and exercise management/information systems to track the course of an engagement.

Many major posts across the country have a local MOUT facility where units conduct urban warfare training during the year in order to prepare for the MOUT phase of their JRTC rotation.[4] The difference between the JRTC Shughart-Gordon MOUT complex and other MOUT facilities across the country is added realism, more experienced and numerous observer/controllers (O/Cs), and the instrumentation and digitized infrastructure that allows more thorough evaluations and after-action reviews. In

---

[4]MOUT facilities exist at Fort Drum, Fort Bragg, Fort Lewis, Fort Campbell, Fort Benning, and the Marine post of Camp Lejeune.

an emergency, perhaps some of the instrumentation could be reduced or eliminated. But if soldiers were being prepared for an urban contingency, as much realism as possible should be incorporated into the urban warfare center.

- **Expansion.** Light force training facilities do not require such a large maneuver space for battalion-level force-on-force exercises as do heavy force training facilities. Light forces have less mobility than heavy forces, so a somewhat smaller maneuver space is adequate. This means that smaller facilities would be sufficient. For example, Fort Irwin is about 600,000 acres; Fort Polk is only about 200,000 acres. If a JRTC-like facility could be built on every Army land tract of 200,000 acres or more, there would be space for a large number of facilities.

Probably the major constraint on the location of these facilities would be the realism of the terrain. Depending on the local geography of a contingency in a particular area of the world, it would be best to locate these facilities in similar terrain. However, because the urban warfare component of the training may be the most important part, the MOUT complex may be the most important feature of the facility. Therefore, the local geography may not be that important.

To achieve the level of realistic training that the JRTC provides, one cannot simply send some observers/controllers (O/Cs) somewhere. An already existing base infrastructure would have to be located and supplemented by villages, towers, radios, etc. One possibility would be to set up shop again at the original JRTC site, Fort Chaffey. O/Cs and an opposition force (OPFOR) would have to be trained, and vehicles for the O/Cs would have to be procured.[5] This process would take about six months at a minimum to begin the first tentative rotations.[6]

The duplication of a light force CTC might be achieved within three to four months if a more primitive "Fort Chaffey style" JRTC with a "bare bones" instrumentation setup was the goal. In

---

[5]Currently, visiting brigades to the JRTC fall in on a prepositioned fleet of about 600 vehicles (trucks, artillery, HMMWVs).

[6]Based on a conversation with COL Pickens, O/C Commander.

this case, a small database and a portable trunk radio system might be sufficient for the minimal infrastructure. Otherwise, instrumentation is clearly the constraining factor. If any lead time for procuring communication or instrumentation technology is involved, then it could take from 12 to 14 months.

We now describe the required personnel to conduct advanced training.

- **Requirements.** The JRTC requires two groups of training personnel: the OPSGROUP and the OPFOR. The OPSGROUP consists of the O/Cs. There are about 660 O/Cs, who "cover down" a visiting line brigade to the squad level. Every O/C must have direct experience in the position being covered. The OPFOR is a dedicated U.S. Army Airborne infantry battalion, the 1st of the 509th, which replicates a hostile force and provides a level of realistic collective training that cannot be duplicated at home stations. There are 440 personnel in the OPFOR and 660 in the OPSGROUP. O/Cs are mostly senior NCOs (E-6s and E-7s) and captains (O-3s). A breakdown of just the infantry instructors in the OPSGROUP is shown in Table 6.1.[7]

  O/Cs must have command experience in the particular position that they would be observing and controlling. Most of these infantry officers and NCOs have served in relatively elite units like the 82nd Airborne and the 101st Air Assault Division. Most are Ranger-qualified.

- **Expansion.** Expanding to more JRTCs would require the use of active component TOE forces that possess the uniquely experienced personnel capable of performing as O/Cs. Although it is difficult to estimate the number of active TOE forces required for trainers and training managers, we have assumed for our nominal estimate that about one-and-one-half active light brigades could provide the trainers necessary to man one light force combat training center.[8]

---

[7]Data are from the OPSGROUP TDA listing.

[8]This is based on an examination of infantry personnel only. Other critical MOSs such as special forces and Civil Affairs are outside the scope of our exploratory modeling approach.

Table 6.1

OPSGROUP Instructors

Enlisted MOS	Pay Grade			
11B, 11M, 11C, 11H, 11Z	E-6	E-7	E-8	E-9
	78	65	12	4
Officer AOC title	Pay Grade			
11A	O-3	O-4	O-5	O-6
	63	17	8	2

Maximum capacity. At the JRTC during peacetime, two months are normally skipped in the rotation cycle to allow the cadre vacation time and to train new O/Cs. In an emergency, these two extra months could be used. Furthermore, a rotation lasts for only about three weeks. If the full capacity of the training facility were used, more units could be trained. However, only two of the three battalions of a brigade are actually trained while the third battalion participates in Task Force simulation. Probably all three would need to undergo actual training in an expansion.

When all these factors are considered, a more precise estimate of the training capacity of a JRTC-like site is difficult to determine. Our nominal estimate is that for one training site, the maximum rate of advanced training is 12 light brigades per year. In our exploratory analysis we will vary the estimate. Each additional JRTC would add a multiple of this after a startup delay.

We have assumed that the number of JRTCs could be expanded in about six months. Figure 6.2 shows the rate of training of light forces at the JRTC level with these nominal values.

If the capacity of each JRTC is 12 brigades per year, then after a six-month delay, three JRTCs could train soldiers at a rate of about 36 brigades per year and six JRTCs at about 72 brigades per year. As we can see from Figure 6.2, using these nominal estimates, 40 divisions of light forces (120 brigades) could complete advanced training with six JRTCs in about two years. Next, we examine the capacity of the ARNG and the basic and MOS training process to determine their potential rate of making units available for JRTC training.

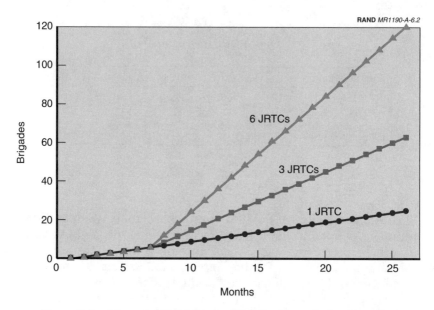

Figure 6.2—Advanced Light Forces Training Cumulative Capacity

Training Up for ARNG Deficiencies

ARNG units usually have a number of deficiencies that must be corrected before they are ready for additional training. The process of correcting these deficiencies for light forces will be similar to the process discussed above for heavy forces (see Chapter Three). We assume that IRRs could fill deficiencies in ARNG units. Just as for heavy forces, our nominal estimate for these units to be ready for advanced training is two months.

New Unit Training

To expand to 40 light divisions, the Army would have to form new units. Newly formed units would require the recruitment of enlisted personnel, NCOs, and officers. In the heavy division section of this report, we discussed the basic training mechanism that produces new infantrymen. As far as the infantry is concerned, that same training model applies here.

Facilities. Light divisions require light infantrymen or foot soldiers, the 11B MOS. Both 11Bs and 11Ms (Bradley infantrymen) are trained at the infantry school at Fort Benning. Unlike the advanced training at a JRTC, the facilities at Fort Benning are not heavily instrumented. Such a facility would require mostly ranges and other low-tech training facilities and could easily be expanded.

Trainers. There are four training units that conduct infantry training at Fort Benning:

- 29th Infantry Regiment
- Ranger Training Brigade
- Infantry Training Brigade
- 11th Training Brigade

We focus on the relevant training units for basic infantrymen, the 29th and the Infantry Training Brigade (ITB).[9]

The 29th Infantry Regiment supports the infrastructure of training ranges, facilities, and any other special sites or equipment and provides the range instructors. These are all necessary to teach infantry skills to basic infantry recruits, junior infantry officers, and Ranger school students.

The Infantry Training Brigade (ITB) is the actual unit that holds the raw recruits and their drill instructors. It currently contains eight training battalions of infantry recruits, not all of which are actively manned and working at any one time, and the drill instructors (DIs) who conduct basic training. There are currently about 400 drill instructors, primarily E-6s and E-7s. To become qualified, they must attend a nine-week drill sergeant school. There are 12 instructors per basic training company (five companies per training battalion).

The most likely constraint on expanding basic and MOS training would be trainers from the reserves who run the ranges and conduct

---

[9]The Ranger Training Brigade comprises the Ranger instructors and students in Ranger School. The 11th Training Brigade contains all the students in Infantry Officer Basic and Advanced courses, the NCO academy, and the instructors associated with these courses.

formal infantry classes. Many of these noncommissioned officers are activated from the reserves and sent to Fort Benning to help out the trainers in the 29th Infantry Regiment. However, our assumption is that in an emergency, trainers from active units would be used. Thus, we do not believe this problem would actually constrain expansion.

TRAINEES

There are currently not enough active and reserve forces to expand light forces to 40 divisions. This expansion would require not only all of the current reserve forces, but new units as well. The various unit types and the training requirements for each are taken up separately, beginning with the active units.

Active Component or "Ready" Brigades

The AC brigades are nominally ready for deployment, although we use them as trainers in our model for a large expansion. There are currently four divisions of light forces—the 82nd, the 101st, the 10th, and the 25th. These forces are all organized, equipped, and trained somewhat differently, but they are all light infantry forces and could be deployed in a contingency, if necessary. They maintain their individual skills as part of their daily routine.

ARNG Brigades

ARNG brigades are the next type of force to be readied. In Figure 6.1, these forces are represented first by ARNG(–), which indicates some deficiencies, and then by ARNG, which represents the same forces with their deficiencies corrected. After correction, these forces are ready for advanced training.

Included in the ARNG are the enhanced ready brigades discussed above. The 15 ARNG brigades designated by the Army as ERBs are heavy brigades. In an emergency, however, we assume that these brigades could be restructured and retrained as light brigades. The infantry in these ERBs would be ready to form new light units. Those trained in armor specialties would require some additional training. But because 90 percent of the personnel in these brigades are present, we assume that these ERBs could be ready to go to a JRTC-like advanced training facility after about eight weeks of additional training.

In addition to the 15 ERBs, the ARNG has eight more heavy divisions (or at least division flags) as of this writing. The C-rating of these eight divisions varies, but, in general, it is considerably below the C-1 ratings of the ERBs. Although these ARNG divisions are heavy divisions, again we assume that after correcting their deficiencies and retraining some specialties, these units could be restructured into light forces and made ready to go to a JRTC-like training facility within about eight weeks.

New Units

Civilians are the last type of force to be readied. Enlisted civilians first receive basic and MOS training, get a complement of NCOs and officers (who might also need training), and emerge as an active component unit. This unit is then ready for training at a JRTC. Newly formed units would require the accession of enlisted personnel, NCOs, and officers.

Enlisted personnel. Training battalions hold trainees who are at different stages of their training cycle (which is 13 weeks for regular 11B infantry and 15 weeks for 11M infantry). Graduation rates can be increased by shortening one station unit training (OSUT) from 13 to 12 weeks.[10] There are currently six infantry basic training battalions of recruits, which equates to about 13,000 infantry graduates per year. Over the past several years, the number of graduates has varied from about 10,000 to about 16,000. But in 1991, 23,000 were trained.

Fort Benning mobilization plans during the 1980s called for an expansion to 27 concurrently running training battalions, which could have produced an annual basic training rate of 78,000. This translates into about 32 brigades per year. Maximum capacity is even higher if we begin to consider alternative infantry training sites other than Fort Benning.

---

[10]OSUT combines basic and advanced infantry (MOS) training into one training phase.

In the Army's World War II experience (see Chapter Three), a large number of enlisted personnel were trained quickly.

Noncommissioned officers. Requirements for NCOs for light forces are parallel to those for heavy forces. Because a light infantry division requires about 500 E-6s, then if six JRTCs were built, about 12,000 infantry E-6s would be required per year.

Just as for heavy forces, NCO schools will probably be able to train enough NCOs to fill all ranks. But in a major expansion, problems could come with the lack of experience of some NCOs. This lack of experience could be important at all ranks, but the greatest problem could come at the E-6 level. Many of these E-6s are squad leaders, a position that requires skill, training, and several years of experience.

Officers. During a national emergency that resulted in a major force expansion, more officers of all grades would be required. However, through promotion and a call-up of the IRR, most senior officer grades could probably be filled. In World War II, the most critical shortage of officers was at the most junior rank—second lieutenants.[11] In another major expansion, O-1s would again probably be most in demand, because many new O-1s would be required and there are no lower ranks from which to promote.

In a light division, we assume that every platoon is led by a second lieutenant, which gives a requirement of about 160 infantry platoon leaders required per division.[12] If 12 to 24 divisions could undergo JRTC-like advanced training every year, then about 2,000 to 4,000 infantry second lieutenants would be required, with the same number required each year. During the Vietnam War, well over this number of second lieutenants were produced by the OCS system (see Figure 3.4).

------

[11]This was especially true for the infantry branch. For simplicity, we assume all platoon leader positions are held by second lieutenants (O-1s), even though this command position can also be held by first lieutenants (O-2s).

[12]Here we assume that there are four platoon leaders per company, three companies per battalion, and 10 battalions per light division.

MAXIMUM PRODUCTION RATE

We have now estimated the flow rate for advanced training JRTCs, ARNG units, and new units. Let us now compare these rates to determine how smoothly the flow of trained light forces might be in the nominal case.

In Figure 6.3, we again show the training capacity of one, three, and six JRTCs. The curves represent the final preparation stage before these units are ready for combat. Then we have shown the approximate time required to be ready to attend a JRTC for the AC, ARNG, and new units.

As Figure 6.3 shows, the training rate for light forces at a JRTC is considerably faster than for training heavy forces (compare with Figure 3.2). In fact, with six JRTCs open, the rate of providing basic and MOS training for new units would probably constrain the expandability process after about two years. Nevertheless, according to

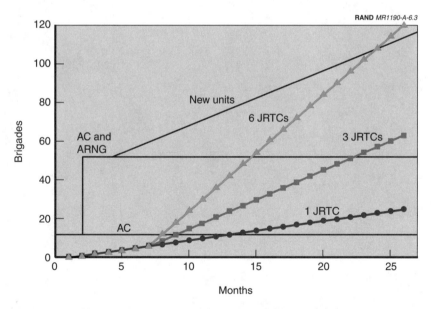

Figure 6.3—JRTC Capacity Versus Unit Availability

these nominal estimates, about 40 divisions of light forces could be trained in about two years. In the exploratory analysis we will see how the time can vary with different assumptions.

Chapter Seven

EXPANDING THE ARMY'S LIGHT BRIGADES IN THE FUTURE

THE MODEL

Figure 7.1 is a flow diagram of the basic model. The parameters used, their nominal values, and the high and low values of their ranges are in Appendix B.

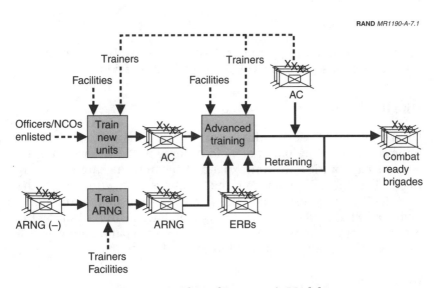

Figure 7.1—Flow of Parametric Model

BOTTLENECKS IN EXPANDING LIGHT FORCES

The bottlenecks in the expansion of light forces are different from those of the heavy forces. One obvious difference is the lack of a serious industrial base component in the expansion of light forces. More important, because of the relative speed with which light forces can be readied, several parameters, in addition to the number of advanced training sites, can come into play. In all, the following parameters can affect the expansion of light forces:

*   Number of JRTCs
*   Advanced training time
*   Individual training capacities
*   Individual training time
*   Initial reserve strength
*   Retraining time
*   Expansion delay

For purposes of sifting through these effects, we have chosen the number of JRTCs as a means of focusing the discussion.

One JRTC Only

Unlike the case for expanding heavy forces, it is difficult, but not impossible, for other parameters to become bottlenecks even if the current capabilities are not expanded beyond the one operating JRTC. The one parameter from the list above that *cannot* affect the one-JRTC case is, of course, the site expansion delay—the delay before additional JRTCs can be brought on line. Figure 7.2 shows the expansion rate of light forces for today's nominal parameter values if there is only one advanced light forces training site.

Observations

*   Any increase in advanced training time for light forces will depress the expansion rate in Figure 7.2 proportionately. Conversely, any decrease will increase it proportionately.

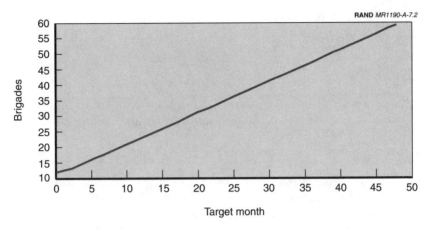

Figure 7.2—Expanding Light Forces with One JRTC

- The rate of advanced training is slow enough that civilian brigades do not become involved in an expansion until the fourth year. At that point, individual training capacity and individual training duration can be much smaller than they are today before having an effect on the expansion rate.

- Reserve strength can become a bottleneck, but only at levels well below those of today. Below 12 initial reserve brigades (with ERBs preferred because of their higher initial readiness), the expansion rate will suffer. Near 12 initial reserve brigades, any degradation in other parameters—including individual and advanced training time and individual training capacity—will depress the expansion rate.

- As with all other cases, if brigades must be retrained, the point at which that occurs for the readiest brigades is the point at which the expansion system becomes clogged with retraining, unless new JRTCs are opened.

Interpretation. If no new JRTCs are built, advanced training for light brigades is more than likely to remain the bottleneck to expansion. Other parameters can become bottlenecks, but generally only at levels that are plausible but unrealistic. If the duration of advanced training increases substantially, those bottlenecks occur at more plausible levels.

Two JRTCs

Building a second JRTC basically doubles the throughput of the light forces expansion system, but only if other elements keep pace.

Observations

- Of the elements that must keep pace, the most important is the individual training capacity. It takes a 25 percent increase in individual training output to keep a four-year expansion from being affected. This, however, depends on having as many reserve brigades as today. Any drop in the total number of reserve brigades affects the light forces expansion rate. This is because the rate of individual training is not commensurate with the rate of advanced training unless the throughput of individual training is doubled. It can keep up with the advanced training capacity of two JRTCs because there is a sufficient backlog of reserve units to train up. (Oddly enough, tripling the individual training output would actually depress the expansion curves slightly because of the need to use ready brigades to man the third site and because the JRTCs then become the bottleneck again.) Figure 7.3 shows this in a different way. It shows nominal parameter values (including for the current individual training capacity) and the light force expansion rates with one, two, three, and four JRTCs. The curve for two combat training centers shows a slight accessions effect after 36 months, and the curves for three and four centers show serious effects.

- Figure 7.4 summarizes the sensitivities of a light forces expansion system with two JRTCs. The red curve in Figure 7.3 is for a system with nominal values except that, as described above, the individual training capacity has been increased by 25 percent. Even with only one additional advanced training center, there is a noticeable sensitivity to doubling the delay (the blue curve in Figure 7.4) in bringing that site on line.

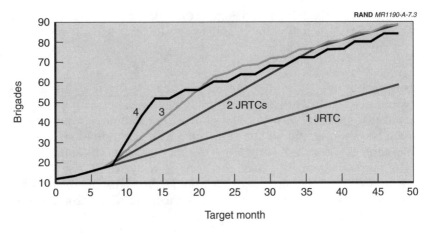

Figure 7.3—The Effect on Expansion of Light Forces of Adding Advanced Training Sites (with other values at nominal levels)

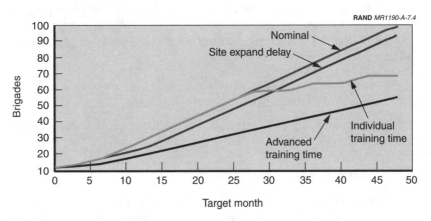

Figure 7.4—The Effect on a Nominal Two-JRTC Expansion Capability (with marginal individual training capacity) of Doubling Other Delays

- Doubling the individual training duration from four to eight months (the green curve in Figure 7.4) causes the expansion to become accessions-limited after the reserves have been through advanced training. Increasing the individual training capacity could alleviate this effect.

- Doubling the advanced training time (the black curve in Figure 7.4) basically halves the overall expansion rate and resembles the one-JRTC case.

Interpretation. Doubling the number of advanced training sites would double the throughput of today's system, but with greatly increased sensitivity to other elements of the system. In particular, such a system would be out of balance with the individual training element. The current individual training capacity, in conjunction with today's reserve brigades, can keep up with two JRTCs. Without those reserves, however, the Army would have to double the individual training capacity to match the advanced training throughput of two JRTCs.

Of the remaining elements of the expansion system, doubling the number of JRTCs would be most sensitive to significant increases in the advanced training time. If that as much as doubled, it would negate the effect of doubling the number of JRTCs. With two JRTCs the expansion system would be sensitive to significant increases in the individual training time, but not until the reserve brigades have all gone through advanced training.

Three or More JRTCs

Adding two or more JRTCs would have effects similar to those for adding just one.

Observations

- Doubling the individual training capacity, in conjunction with today's reserve structure, just about keeps up with three JRTCs, and tripling the individual training capacity just about keeps up with four. As with the two-JRTC case, it actually takes tripling the individual training capacity to keep up with the throughput of three JRTCs, and similarly for four JRTCs.

- Figures 7.5 and 7.6 summarize the sensitivities of a light forces expansion system with three and four JRTCs, respectively. The red curve in each figure is for a system with nominal values except that the individual training capacity has been increased to

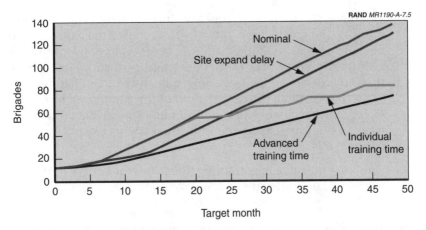

Figure 7.5—The Effect on a Nominal Three-JRTC Expansion Capability (with marginal individual training capacity) of Doubling Other Delays

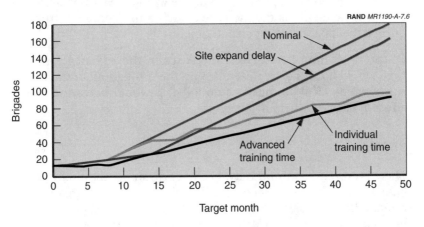

Figure 7.6—The Effect on a Nominal Four-JRTC Expansion Capability (with marginal individual training capacity) of Doubling Other Delays

just meet a smooth four-year expansion (similar to the 25 percent increase for the two-JRTC case). Once again there is a noticeable effect from doubling the delay in bringing extra sites on line (the blue curve in the figures).

- Doubling the individual training duration from four to eight months (the green curve in the figures) causes the expansion to become accessions-limited after the reserves have been through advanced training. For the greater number of additional JRTCs, the reserves are trained up more quickly, so this effect occurs earlier. Because of the earlier onset of this limitation, its overall effect almost matches that for doubling the advanced training time. Again, increasing the individual training capacity could alleviate this effect.

- Doubling the advanced training time (the black curve in the figures) basically halves the overall expansion rate.

Interpretation. As long as the individual training capacity is increased by a factor one less than that for the JRTCs, today's reserve structure would keep a light expansion "balanced" for four years. In general, the individual training capacity would have to be increased by the same factor as the increase in JRTCs to maintain a true balance in the light forces expansion pipeline.

Of the remaining parameters, increases in the advanced training time would have the greatest effect on the throughput of the expansion system. If it as much as doubled, it would halve the overall effect of the increase in JRTCs.

SUMMARY

There is currently only one advanced or brigade-level training site for light forces (the Joint Readiness Training Center). If this were to remain the only such site during an expansion, that expansion would be primarily limited by the throughput there. Any increase (decrease) in the advanced training time would proportionately decrease (increase) the expansion rate. Other elements of the light force expansion system could become bottlenecks, but only under rather extreme conditions.

With a second advanced training site, the expansion capability is much more sensitive to the individual training capacity. In general, if the individual training capacity doesn't keep pace with the increase in JRTCs (through a failure to increase the number of individual training sites, through an increase in the individual training time, or through a failure to access enough recruits to keep the individual training pipeline full), individual training can become the bottleneck to the light force expansion system. A large reserve structure, such as today's, can alleviate a problem with the individual training capacity, but only until those reserves are trained up. The greater the number of advanced training sites, the faster the process of training up the reserves.

As earlier, this sensitivity to individual training is a surrogate for a sensitivity to attracting recruits for a volunteer Army. Expanding beyond today's 57 ready and reserve brigades (or a different total in the future) will rest on the ability of the recruiting system to keep the expansion pipeline full. With the wide variety of possible light force missions, this could be particularly problematic. Without assuming the timely reinstitution of a draft, the process of accessing volunteers must be considered a particularly likely (and unpredictable) bottleneck to any light forces expansion that goes beyond the ready and reserve forces in place.

The reserve forces play a more sensitive role in light force expansions than they did in the heavy force expansions. The light force expansion system generally "uses up" reserve forces more quickly than the heavy force expansion system. This brings variables other than individual training into play and makes the entire system particularly sensitive to how many reserve forces there are in waiting.

Retraining time plays the same role in light force expansion as in heavy force expansion. If it must occur, it will completely clog the expansion pipeline unless new advanced training sites are opened.

Chapter Eight

COSTS AND EXPANDING THE ARMY'S LIGHT FORCES IN THE FUTURE

INTRODUCTION

As with the heavy costing analysis, the costing analysis for light force expansion was focused on two narrow, but general, questions:

- What can be said (both now and in the future) about the Army's ability to expand to a given light force size in a given time that minimizes recurring peacetime costs?

- What can be said (both now and in the future) about the Army's ability to expand to a given light force size in a given time that pays attention both to recurring peacetime costs and expansion costs?

In answering the first question, the research involved looking directly at a comparison of the recurring peacetime costs of a light force expansion capability with the time required to expand given that capability. Recurring costs are as defined in Chapter Five.

In answering the second question, the research was again done in expansion size-time state space in which each point represents a target size and a target time to reach that size. A minimal-recurring-cost light force expansion capability was computed for a subset of points in size-time space, allowing for an exploration of the resulting force structures. Of particular interest in these explorations were the expansion costs associated with the minimal-recurring-cost expansion capabilities. Expansion costs are as described in Chapter Five.

The model is the same one described in Chapter Seven, and the costing numbers used for the model are documented in Appendix B.

RECURRING COSTS VERSUS THE TIME TO EXPAND LIGHT FORCES

Again, for the costing explorations we were interested in points in expansion size-time space—that is, we were interested in the recurring costs of expanding where there is a target size for the expansion and a target time allowed in which to complete the expansion. In computing the recurring costs, the exploratory model allows for any of the model parameters to be set beforehand except for the initial number of ready brigades. It then determines the number of initial ready brigades required to reach the target size by the target time, returns that as an output, and computes the recurring costs that include the required ready brigades.

For the most part the exploration of recurring costs for light force expansions, too, was uninteresting in the sense that there were few surprises. Changing cost parameters generally changed the overall costs in a "smooth" way.

As with the heavy forces case, there are "kinks" in the recurring cost curves when—as with the nominal case—the recurring costs of a ready brigade are significantly larger than those of an ERB, a NG brigade, or a brigade of civilians.

Observations. Figure 8.1 shows the recurring costs for reaching targets of 15, 30, 45, and 60 brigade targets. The model parameter values are all at nominal levels (except that the maximum number of JRTCs that could be built during the expansion was set to three). Figure 8.1 shows much the same character as the corresponding graph for the heavy forces case.

The nominal recurring cost of a light ready brigade is almost four times that of an ERB and almost five times that of a NG brigade. Lowering those ratios "flattens" the curves in Figure 8.1. Even at equal costs for ready, ERB, and NG brigades, a slight knee exists when civilian brigades are brought to readiness.

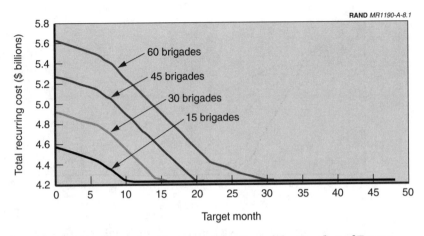

Figure 8.1—Recurring Costs as a Function of the Number of Target
Brigades and Target Month (for three JRTCs)

The parameter that most affects the drop in costs is, again, the number of JRTCs. The more there are, the more quickly reserve brigades can be trained up. This reduces the number of ready brigades that would need to be maintained to reach expansion target sizes, thereby reducing the recurring costs more quickly. Figure 8.2 shows the effects of the number of JRTCs brought on line during expansion. At target times from 11 to 15 months, the drops in recurring costs continue all the way through six JRTCs.

The remaining variables can change the recurring costs of an expansion capability for light forces up or down, but they do not measurably affect the basic shape of the cost curves.

Interpretation. The longer the Army has to expand the light forces, the lower the recurring costs required to maintain the ability to expand. For a given target expansion size, there is an expansion time at which ready troops would not be required (except, possibly, as trainers for the reserve or civilian brigades) in order to meet the target size on time.

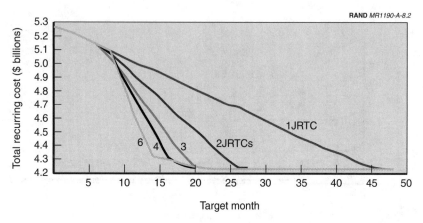

Figure 8.2—Recurring Costs as a Function of the Number of JRTCs and Target Month

Recurring costs drop with more JRTCs, because the increased training capacity means the number of ready troops required to meet size targets is reduced because of the ability to train reserve and civilian troops more quickly. If there are fewer reserve troops and the individual training capacities are at or less than nominal, meeting size targets will require more ready troops for longer periods, pushing the knee out farther in time and lessening its sharpness. Cost savings stop when there is enough time to meet the target entirely with civilian units.

These findings are generally unsurprising and rest on the disparity in recurring costs between ready brigades and reserve brigades.

RECURRING AND EXPANSION COSTS

For this part of the analysis, we will again look at recurring and expansion costs by the number of JRTCs, but again with the number of JRTCs being the maximum number allowable in an expansion. That is, the exploratory model looked for the optimal force structure and expansion system that minimized costs and may not have chosen to build all the JRTCs allowable. The JRTC value is the total number it could build.

Three JRTCs

Observations. Figure 8.3 shows the minimal recurring costs in size-time space. These are calculated for the centroids of the boxes in a 50×50 grid.

Again, these are not the full recurring costs of the expansion capability. It is not the costs per se that we are interested in, but the shapes of the regions and how they change with changing circumstances. These graphs are, however, a good representation of the change in recurring costs as the target size and allowable time change. Again, if a large Army is required quickly, the recurring costs rise rather steeply as the target size increases. The area of steepness is smaller than for heavy force expansion because of the shorter advanced training time required to ready light forces.

As in the heavy forces case, the recurring costs are driven primarily by the number of ready brigades required to meet the expansion target on time, as can be seen in Figure 8.4.

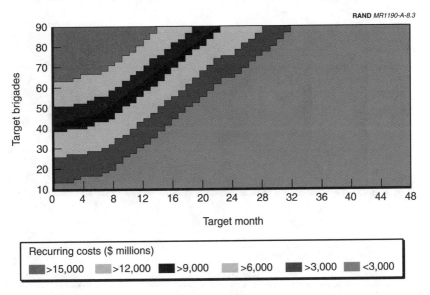

Figure 8.3—Minimal Recurring Costs with No More Than Three Allowable JRTCs

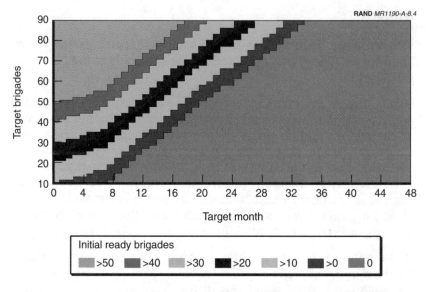

Figure 8.4—Number of Initial Ready Brigades for Each of the Minimal Recurring Cost Force Structures (with no more than three allowable JRTCs)

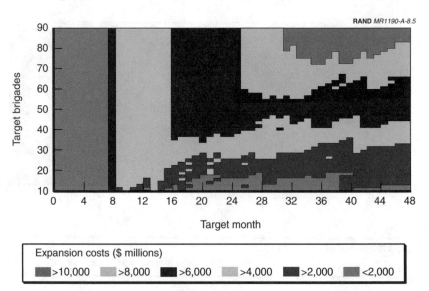

Figure 8.5—Expansion Costs for the Minimal Recurring Cost Force Structures (with no more than three allowable JRTCs)

The expansion costs associated with the minimal-recurring-cost force structures of Figure 8.3 are shown in Figure 8.5. As in the heavy case, there are two major themes in this figure. The first centers around the vertical stripes. They represent the fact that trained-up reserve and civilian brigades are carried as an expansion cost. As a function of time (and independent of other factors), reserve and civilian brigades can be readied and begin adding to expansion costs. The vertical stripes represent regions of these costs.

The second theme centers around the general trend of slowly growing regions of reduced costs as time for expansion gets large enough. The serrations in this light forces plot are more pronounced than those in the heavy forces plots, but again, this represents the interaction between the increase in expansion costs due to increased numbers of readied reserve and civilian brigades and the reduced costs from needing fewer JRTCs to meet the size goal.[1] Figure 8.6 shows the initial number of reserve brigades in each of the minimal-recurring-cost force structures. The banding evident in Figure 8.5 is again evident here, as is the slowly reduced need even for reserve units as less expensive civilian units can be readied.[2]

Figure 8.7 shows the number of JRTCs in each of the optimal force structures. It depicts the decreasing need for additional JRTCs to meet the size-time goals. The uneven conjunction of this with the reduction in the required number of reserve brigades leads to the interesting patterns in Figure 8.5.

Interpretation. Over most of the expansion size-time space, recurring costs and expansion costs vary smoothly. Again, there are no sharp breaks in the expansion costs related to optimal-recurring-cost force structures. The expansion cost graph shows more interesting characteristics than in the heavy forces case. There may actually be some regions worth exploring in a more detailed way to explore the

---

[1]And again, there is an interaction from the optimization routine that finds regions of equal cost alternatives and arbitrarily chooses one to represent the force structure.

[2]The colored "islands" in some of the regions of Figure 5.7 are due to interactions between discretization of the expansion process, the 50×50 discretization of the size-time space, and the optimization routine that computed the minimum-cost solutions. While adding interest to the picture, they do not materially affect the interpretation or conclusions.

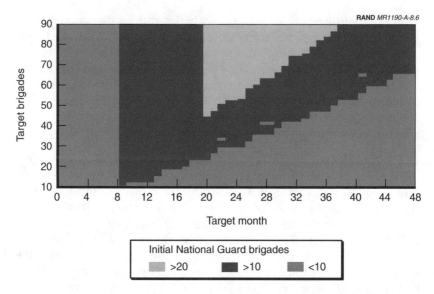

Figure 8.6—Initial Reserve Brigades for Each of the Minimal Recurring
Cost Force Structures (with no more than three allowable JRTCs)

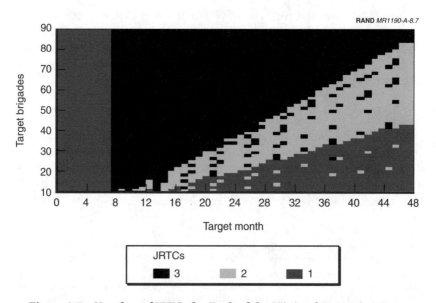

Figure 8.7—Number of JRTCs for Each of the Minimal Recurring Cost
Force Structures (with no more than three allowable JRTCs)

model's suggestion that there is a subtle interplay between the timing of bringing JRTCs on line and the time (and costs) required to bring brigades to readiness.

As in the heavy case, recurring costs are highest when large forces are required in a short period, while expansion costs generally increase with the time of the expansion. Both recurring and expansion costs can be kept low for small, slow expansions.

Six JRTCs

Observations. Figure 8.8 shows the recurring costs of an expansion system that can build up to six JRTCs. In comparison with Figure 8.3, the bands for the six-JRTC case rise more steeply as a function of the target size (and thus drop off somewhat more precipitously as a function of time).

Figure 8.9 shows the number of ready brigades required to meet expansion targets on time for the six-JRTC case. Again, it drops off more quickly as a function of time.

Figure 8.10 shows the expansion costs associated with the minimal-recurring-cost force structures of Figure 8.10. While it displays much the same character as the expansion costs in Figure 8.5, there are even more interesting regions than earlier. Again, there is a basic difference in how quickly the bands rise as the five new JRTCs come on line and new brigades are readied at a faster rate. Figure 8.11 shows the number of JRTCs used by each of the optimal force structures of Figure 8.8. It is identical to Figure 8.7 for short-time expansions or small expansions. Where the expansions are larger and more time is available, bringing on more JRTCs allows for smaller recurring costs while still meeting the time target.

The even more interesting striping and island patterns in Figures 8.10 and 8.11 are certainly partly a function of the timing of when brigades enter advanced training. There may, however, be something else going on here that is worth exploring in a more detailed way offline. Again, the model is suggesting that if the sequencing of training is just right, one could optimally meet an expansion goal with six JRTCs, while a slightly larger goal could be met in less time with the same sequencing and only five JRTCs. The behavior is not

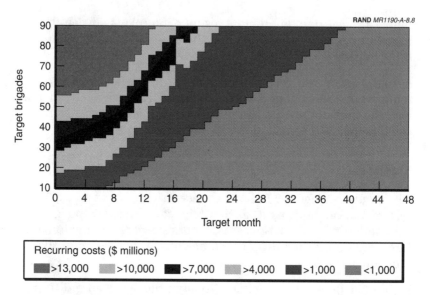

Figure 8.8—Minimal Recurring Costs with No More Than Six
Allowable JRTCs

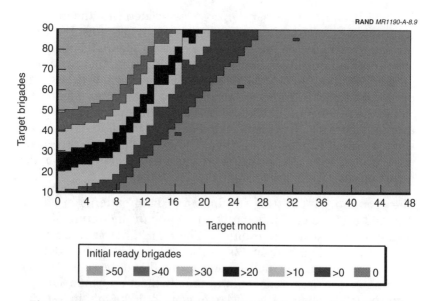

Figure 8.9—Number of Initial Ready Brigades for Each of the Minimal
Recurring Cost Force Structures (with no more than six allowable JRTCs)

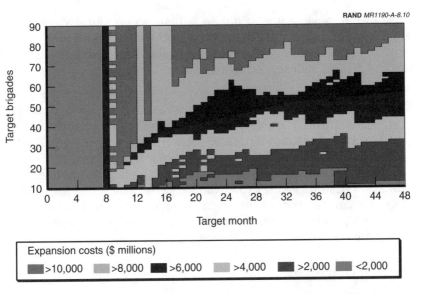

Figure 8.10—Expansion Costs for Each of the Minimal Recurring Cost
Force Structures (with no more than six allowable JRTCs)

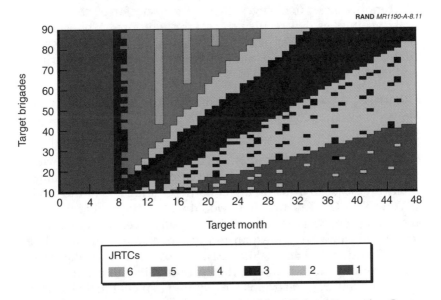

Figure 8.11—Number of JRTCs for Each of the Minimal Recurring Cost
Force Structures (with no more than six allowable JRTCs)

real in the sense that it would be difficult to take advantage of this fact in planning because of the great uncertainties both in the target size of an expansion and in the time available to complete it. In this case, however, it deserves further exploration.

Interpretation. Recurring and expansion costs for short-time contingencies are unaffected by the number of JRTCs that might be built during an expansion. There are large regions in the small, slow expansion area where the maximal number of JRTCs that can help an expansion is low. This obviates the need for a large number of additional JRTCs during an expansion for those cases. Adding several JRTCs, then, can reduce the minimal-recurring-cost force structure required to meet expansion goals, but only for expansions whose goals are long enough and large enough.

CONCLUSIONS

In today's Army, light forces can be trained up more quickly than heavy forces. This causes the cost curves for light force expansions to react more quickly to changes than in the heavy forces case. In general, this implies that recurring costs for light force expansions can be kept smaller than for heavy force expansions. The one exception to this picture today is the nominally longer time it takes to bring on additional JRTCs, compared to NTCs.

Otherwise, there is little surprise in the costs of light force expansions. They tend to vary smoothly with changes in expansion conditions.

It is unlikely that more than one JRTC (the one at Fort Polk) will be maintained during peacetime. In the case of light forces, it would be relatively easier to bring on a second JRTC at the site of the original JRTC at Fort Chaffey, but there would still be the problem of manning and procuring the vehicles for that site. Even so, for periods of less than half a year, the expansion system cannot add much in the way of readied brigades. Similarly, for expansions smaller than a half-dozen brigades per year, the current ability to expand is adequate and only reserve troops are required to meet timelines. It is in between these extremes that an expansion capability beyond today's normal training capability can add cost-effective expandability. The area is smaller than that for heavy forces expansion because of the shorter timelines in general.

As long as the recurring costs of a ready brigade are significantly greater than those of a reserve or civilian brigade, the time and ability to ready a reserve or civilian brigade would permit a substantial reduction in the recurring cost required to provide the expansion capability.[3]

Light forces are readied in large discrete units. Because of this, the timing of when brigades come out of advanced training at several training sites can make a difference when trying to compute the optimal number of training sites to bring on line. In this case there may be situations in which one could take advantage of some of these real differences. It would take a more detailed analysis to prove or disprove this point.

----

[3]Again, this is the "accountant's assumption" that reserve or civilian brigades brought to readiness are the equivalent of ready brigades. That may be closer to accurate in the light forces case if the actual situation is one that the ready forces have done little training for as well, but it is an important caveat in all discussions of expandability.

Chapter Nine

CONCLUSIONS

The analysis of expandability was divided along three primary dimensions:

- Current versus future expansion

- Heavy versus light expansion

- Expansion timing only versus expansion timing and cost

Current expansion was studied to develop a simplified model of expansion in order to do an exploratory analysis of expansion possibilities in the future.

The remaining two dimensions produce four distinct cases for discussion: timing and costs for each of heavy and light force expansions. Each will be taken in turn, to be followed by some general concluding remarks and recommendations.

HEAVY FORCE EXPANSION TIMING

The expansion of heavy forces is dominated by a logic that says the primary impediment to expansion today is training, particularly advanced brigade-level and division-level training, and that after there are sufficient trained brigades to man current equipment, the primary impediment would become the ability of the industrial base to produce more equipment. **Few changes in the future are likely to upset this logic.** Even some drastic changes in the number of initial ready, enhanced ready, or National Guard brigades won't affect it. Small or moderate changes in the duration of individual or advanced training or in the preparation times for enhanced ready or National Guard brigades won't affect it. **Building several new NTCs would hasten the point at which the industrial base becomes the bottleneck, but this wouldn't change the fact that advanced training will be the bottleneck at the start of any expansion.**

The few changes that would affect this logic are:

- In expansions beyond current reserve forces, recruiting (or conscription) must keep current individual training sites full to keep up with three NTCs and must be expanded to keep up with four or more. Said another way, recruiting and individual training can become bottlenecks in any expansion that taps the civilian population.

- If ready troops require retraining, as soon as the retraining begins, the system will be unable to produce additional ready brigades unless additional NTCs are brought on line. That is, the expansion system can be brought to a complete halt by retraining.

- **If current equipment inventories were to suffer severe neglect or become militarily obsolete, the industrial base could quickly overtake advanced training as the bottleneck in the expansion system.**

On the other hand, these changes should be generally visible beforehand. That fact leads to some recommendations below.

In addition to questions about the first-order logic of expandability, there are more general concerns about the Army's ability to expand the training system in the future:

- Today there are sufficient trainers to man three NTCs. If more NTCs were to be built, ready units could be used as trainers at the additional NTCs (at a cost in overall readiness). In the future, the current trainer base is likely to erode and could do so to the point it would be difficult to man as many NTCs as the Army might want to add.

- There is a continual worry about leadership. Primary among these is the worry about having enough E-6s in the NCO ranks and second lieutenants in the officer corps. These worries have been overcome in previous expansions (most notably in World War II), but they remain a concern for any large expansion because of the increasing complexity of the responsibilities that Army leaders are asked to shoulder.

HEAVY FORCE EXPANSION TIMING AND COST

Because cost is important during peacetime and time is most important during an expansion, the primary tradeoff of interest is between recurring (peacetime) costs of expansion capabilities and the time required to expand starting with those capabilities. We looked at two different aspects of this tradeoff: (1) we looked directly at the tradeoff between peacetime recurring costs of an expansion capability and its ability to perform during an actual expansion, and (2) we looked at "optimal" expansion capabilities (for reaching a given force level within a given time) and the effect those capabilities had on expansion costs.

Recurring Costs

We included only the recurring costs of the units and facilities that would be used during an expansion. This included the recurring costs of those ready and reserve units that would actually be trained during the expansion plus the recurring costs of the facilities (such as the NTC) that would be used and were being maintained before the expansion began. It also included the recurring costs of ready units that would be needed to man any additional training sites brought on line after the expansion began. Since the primary objective was comparative costs under a variety of conditions, the industrial base—which is likely to remain in a constant "warm" condition prior to an expansion—was dropped as a recurring cost.

There are no particular surprises in the costing arena.

- For short expansion times (less than a year or so), most of the Army's required capabilities must be in high readiness, with a concomitant increase in recurring costs.

- For expansions that require fewer than six or so brigades per year, the current ability to expand is adequate, and only reserve or civilian units would be required to meet expansion timelines.

- As long as the recurring costs of a ready unit are substantially greater than those of reserve or civilian units, the number of initial ready brigades required to meet expansion size and time targets will drive recurring costs.

- Expansion times between about 6 and 30 months see the greatest benefit in reduced recurring costs when additional NTCs are brought on line during an expansion.

Recurring and Expansion Costs

If one were to know in any expansion exactly how large the Army should become and how much time it had to reach that size, one could compute the minimum-recurring-cost force structure required to accomplish that expansion. Computing such a force structure for a variety of size and time combinations provides some insight into the expansion costs and parameters associated with each expansion. In this case, expansion costs are basically those costs during expansion that wouldn't have occurred if the expansion hadn't taken place. This includes the incremental costs of units that have been brought to readiness as well as the capital and operating costs of bringing new training facilities on line during the expansion.

- Minimum recurring costs vary smoothly over size-time space.

- Expansion costs exhibit two competing characteristics: (1) the general increase in costs due solely to the increased time that trained-up reserve and civilian units must be maintained, and (2) a slow decrease in costs as the expansion time allowed increases is due to a decreased need over time to maintain reserve units or bring new NTCs on line in order to meet size and time targets.

- Where recurring costs are highest, expansion costs tend to be lowest (as there basically *is* no expansion possible because of the short timelines).

- Long, small expansions require the lowest recurring and expansion costs.

- Long, large expansions incur large expansion costs but low recurring costs.

- Large expansions in the (roughly) 12- to 30-month time range are the most costly.

LIGHT FORCE EXPANSION TIMING

Without the requirement for a significant complement of heavy equipment,[1] **the expansion timing of light forces will be driven primarily by advanced training** at a facility like the Joint Readiness Training Center. In contrast to heavy force expansions, the time required to complete JRTC training is short enough that other factors could come into play more easily in the future.

- If only the JRTC is used for advanced training, its throughput will be the expansion bottleneck under all but extreme circumstances.

- The number of JRTCs is again a primary driver of decreased expansion time. With two or more JRTCs, several other factors can become bottlenecks. The most sensitive are the individual training system and the advanced training time.

- **In expansions beyond current or future reserve forces, there must be an increase in recruitment or conscription and individual training sites for each increase in advanced training sites in order to keep recruiting and individual training from becoming bottlenecks.**

- Retraining needs can, again, bring to a halt the production of new ready light units (unless the number of JRTCs is expanded further).

LIGHT FORCE EXPANSION TIMING AND COST

As with heavy forces, there were no particular surprises in the costing explorations. In the light forces case, however, expansion can take place more quickly.

---

[1]The one possible industrial base problem here is trucks. This point was discussed in footnote 24 of Chapter Three.

Recurring Costs

- Recurring costs for short (in this case, half a year or less) expansion times are high and in direct proportion to the target expansion size.

- Expansions as fast as four divisions per year can be accomplished with today's facilities, and only reserve and civilian brigades would be needed to meet the timelines.

- Expansion timelines between (roughly) 6 and 24 months see the greatest benefit in reduced recurring costs when additional JRTCs are brought on line during expansion.

Recurring and Expansion Costs

- Minimum recurring costs vary smoothly (though more quickly than for heavy force expansion) over size-time space and are driven primarily by the number of initial ready brigades required to meet size and time targets.

- Expansion costs again exhibit two competing characteristics: (1) the general increase in costs due solely to the increased time that trained-up reserve and civilian units must be maintained, and (2) a slow decrease in costs due to a decreased need over time to maintain reserve units or bring new NTCs on line to meet size and time targets. In the light forces case, these competing characteristics produce a more varied picture across size-time space.

- Where recurring costs are highest, expansion costs tend to be lowest.

- Long, small expansions require the lowest recurring and expansion costs.

- Long, large expansions incur large expansion costs but low recurring costs.

- Large expansions in the (roughly) 6- to 24-month time range are the most costly.

GENERAL CONCLUSIONS

Expansion depends primarily on bringing people into the Army, training them, and equipping them. There are two general characteristics one would wish for an expansion system. The most important characteristic is that one would want the "pipes" in the expansion system to be "short" and "wide" enough to produce an expanded Army in time. The shorter the time allowed, the less able any expansion system will be to comply, and the more the Army must rely on ready units. In today's Army the expansion pipeline is quite "wide" in parts because of large reserves of both trained soldiers and equipment. This means the primary constraint on expansion today (and into many plausible futures) is training—specifically advanced training at the brigade and division level. If training could be shortened (e.g., by reducing the training time required) and/or widened (e.g., by increasing the number of training sites), today's expansion capability could be improved. Not only is that unlikely in today's climate, those characteristics of today's system are likely to degrade with time (although, if the equipment does not become militarily obsolete, it will remain functional throughout the 15- to 20-year period of study).

The other desirable—but less important—characteristic in an expansion system is that it be balanced, in the sense that there are no serious bottlenecks. **The clear imbalance today in the heavy force expansion system is training**, but that is due to the fact that the Army maintains a large reserve structure ready to be trained (which masks any imbalance in the recruiting and individual training part of the pipeline) and due to the Reagan buildup of equipment (which masks any imbalance in the industrial base part of the system). The heavy expansion system would require three NTCs in order to keep up with the steady-state production out of the Army's individual training sites. For expansions beyond existing equipment levels (currently about 26 divisions of heavy equipment) the expansion system would require an industrial base with two plants operating 2.5 work shifts (the practical maximum) to stay in balance with the remainder of the system.

The light force expansion system, by contrast, **is well balanced**. As fast as people can be brought into the expansion pipeline, they can be moved through it and out the other end. Any increase in the training capability must be accompanied by an increase in recruiting and individual training to keep the light force expansion pipeline balanced.

There are no big costing surprises. Quick expansions will require the large recurring costs of a sizable standing army. The smaller the expansion or the longer the time available for the expansion, the less costly the standing army and its recurring costs have to be. The costs of expansion (unless new equipment is required) are not generally out of line with recurring costs.

RECOMMENDATIONS

The need to expand today's Army is small. The perceived need to be able to expand the Army any time soon is also small. Even so, this analysis suggests some low-cost actions that would enhance the Army's ability to expand in the future. These include both things to do and things to watch for.

Things to Do

- **The biggest positive effect on either heavy or light expansion capabilities would be to decrease the advanced training time.** If this can be done without decreasing military capability, it is the easiest means for decreasing expansion timelines both today and into most plausible futures.

- In the event of a sizable heavy expansion, it would take three NTCs to provide a balanced expansion capability both with today's (and most future) individual training capabilities and capacities as well as equipment stocks. Both heavy and light expansions of a half-year or longer would benefit from further combat training centers. The Army should prepare and maintain plans for building additional centers (and individual training sites if required). This should particularly include where and how they would be built and who would man them.[2]

---

[2]If heavy required forces were to be deployed in or near Europe, the current facilities at Hohenfelds in Germany could be used. For light forces, Fort Chaffey—the original JRTC site—could most easily and quickly be developed into an additional advanced training site.

- Whether or not there is a coming "revolution in military affairs," the best preparation for a large future expansion would be for the Army to maintain and upgrade its current large equipment inventory. As the National Defense Panel's report on transforming defense put it, "It is more important to have a weapon on hand in adequate quantities than to have the capability available to produce that weapon six months or a year later."[3]

- In any large expansion, recruiting could become a bottleneck to the expansion system. A clear and historically employed solution for this is to conscript people into the Army. That remains a viable solution, and it behooves today's all-volunteer Army to retain the ability to implement a conscription system on short notice.

- Because of the shorter time to train light forces, the exact time to train up light brigades is sensitive to the number and phasing of JRTCs that are available for training. A more careful study of the effects of costs and timing in bringing JRTCs on line would be worthwhile.

Things to Watch For

Many of the things to watch for are generally self-evident and currently being done. Nonetheless, it is useful to detail them as a means of reinforcing their connection with the Army's ability to expand in the future.

- Any developments that would make obsolete our current equipment would seriously degrade our expandability capabilities (if not our then-current forces). Such an occurrence would signal the need to reevaluate our industrial base policy (perhaps suggesting a change in its readiness status). Any serious decline in the functionality of the current equipment stocks would also degrade the expandability capability, bringing the much "longer" and "narrower" industrial base into play.

---

[3]*Transforming Defense: National Security in the 21st Century,* Report of the National Defense Panel, December 1997.

- **Any significant changes in training time (positive or negative) at the NTC or JRTC would affect expandability capabilities (negatively and positively, respectively).** Training times should be adequate for proper preparation of Army troops, but changes in those times should be factored into the Army's expansion capabilities, in case changes need to be made there.

- The Army generally does a good job of recruiting for its manpower needs. It also pays close attention to changes in the willingness in U.S. society to serve in the Army. That should continue and be connected into thinking about expansion capabilities. Any expansion that goes beyond then-current reserve forces will be directly dependent on the willingness of young people to serve.

- The final signpost is the most obvious. Because of the critical dependency of having ready troops for military situations that provide short warning time, the Army should look for any hint of a need to expand in size. Any expansion, particularly in tight budgetary times, is likely to be dominated more by political than military concerns. If the Army is ever to expand again, it should begin making the case for expansion as soon as arguable indications of that need arise.

There are those who argue that threats already exist that stress the Army's current ready capabilities. Indeed, the ready military capabilities required to ensure national safety will always be debatable. The purpose of this study has been to give the Army a framework and preliminary guidelines for addressing questions of when, how, and by how much the Army should expand to meet demands that exceed its current capabilities now and into the future.

EXPLORATORY MODELING

Exploratory modeling provides an alternative rationale for using models to understand complex systems. In exploratory modeling, results of a model run are not viewed as a prediction of what we would expect to occur, but are rather the results of a computational experiment. That experiment tells us what the outcome would be if all the guesses we had to make in setting up the model turned out to be true. By making different guesses, different modeling experiments are produced. As any given experiment is based on a number of such guesses, our knowledge about the problem being studied cannot be captured by any single model or experiment. Instead, the available knowledge is viewed as being contained in the collection of all possible modeling experiments that are plausible given what we know.

An important point here is that this style of analysis requires only that the models used are plausible: that they are consistent with what is known. This is a much less stringent requirement than predictivity. Exploratory modeling searches for a viable basis for a policy decision in spite of the uncertainties and unpredictability of the problem. In effect, exploratory modeling looks for a question that can be answered even in the presence of uncertainty or unpredictability. This contrasts to the goal of sensitivity analysis, which is to attach a variance estimate to the predictions of a model.

At its most general, exploratory modeling can be understood as search or sampling over a set of models that are plausible given *a priori* knowledge or that are otherwise of interest. This set may often be large or infinite in size. Consequently, the central challenge of

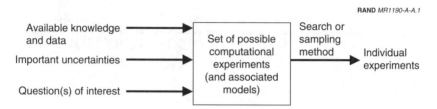

Figure A.1—The Central Challenge of Exploratory Modeling

exploratory modeling (as represented in Figure A.1) is the design of search or sampling strategies that support valid conclusions or reliable insights based on a limited number of computational experiments.

A wide range of research strategies is possible, including structured case generation by Monte Carlo or factorial experimental design methods, search for extremal points of cost functions, sampling methods that search for regions of "model space" with qualitatively different behavior, or combining human insight and reasoning with formal sampling mechanisms. Computational experiments can be used to examine ranges of possible outcomes, to suggest hypotheses to explain puzzling data, to discover significant phases, classes, or thresholds among the ensemble of plausible models, or to support reasoning based upon an analysis of risks, opportunities, or scenarios. Exploration can be over both real valued parameters and non-parametric uncertainty such as that between different graph structures, functions, or problem formulations.

Aggressive exploitation of exploratory modeling for complex models requires significant computational resources. Consequently, this approach has had widespread use only recently. As computer power continues to grow, this approach can be expected to become increasingly important. Exploratory modeling is supported by a set of software tools that facilitate the parallel processing of different problem formulations on a distributed network of computer workstations, exploiting otherwise idle processor time, providing supercomputer levels of processing power without use of a supercomputer. In previous research at RAND, between 10,000 and 500,000,000 computational experiments have been run on networks of workstations (Sun

Sparc 2's, 10's, and 20's) over periods of days to weeks in support of assorted policy analyses.

An Exploratory Modeling Approach to Expandability

There are three general types of applications where exploratory modeling can be used: data driven, question driven, and model driven. Data-driven exploration starts with a dataset and attempts to derive insight from it by searching over an ensemble of models to find those that are consistent with the data. Question-driven exploration begins with a question we wish to answer and addresses this question by searching over an ensemble of models and cases believed to be plausible in order to inform the answer. Model-driven exploration involves neither a fixed dataset nor a particular question or policy choice, but rather is a theoretical investigation into the properties of a class of models, and is consequently a branch of experimental mathematics. Examples of data-driven and model-driven exploration can be found in all the sciences.

Question-driven exploration, on the other hand, has particular salience for our expandability research. The basic question that motivates the exploration of outcomes is: "What changes would be required before there are fundamental changes in the major drivers of expandability?" By looking across plausible future outcomes, we hope to discover those changes in the world or in policy related to expandability that would cause the first-order logic of today's expandability thinking to change. For example, if today a major driver of expandability timelines is the bottleneck at brigade-level training, what changes (and what rough magnitude of change) would be required to cause this to cease to be a major driver?

For further information on exploratory modeling, see:

Steven C. Bankes and J. J. Gillogly, *Exploratory Modeling: Search Through Spaces of Computational Experiments*, Santa Monica, CA: RAND, RP-345, 1994.

Steven Bankes, "Computational Experiments and Exploratory Modeling," *CHANCE*, Vol. 7, No. 1, 1994, pp. 50–57.

Arthur Brooks, S. C. Bankes, and B. E. Bennett, *Weapon Mix and Exploratory Analysis: A Case Study,* Santa Monica, CA: RAND, DB-216/2-AF, 1997.

R. J. Lempert, M. E. Schlesinger, and S. C. Bankes, "When We Don't Know the Costs or the Benefits: Adaptive Strategies for Abating Climate Change," *Climatic Change,* No. 33, 1996.

EXPLORATORY MODELING PARAMETER VALUES

The nominal values for timing and cost parameters of the models represent estimates based on a good deal of research. Our intent was to get them approximately correct. In some cases (such as the costs for building a new combat training center), they represent little more than judgments from Army sources familiar with the costing data. While our intent was to be reasonably accurate, their accuracy is not critical to the analysis we have done. More important than the nominal values are the ranges for each of the parameters. Here we were careful to make the ranges large enough to include any plausible value—current or future—of that parameter. If the ranges encompass all reasonable values, we will have gauged the effects of those values during our exploratory excursions.

HEAVY FORCE EXPANSION TIMING

Table B.1 gives the names, descriptions, and values for the parameters in the model used to explore the timing of heavy force expansion. These values were derived from the research documented in Chapter Three.

Table B.1

Heavy Force Timing Parameters

Parameter	Parameter Description	Nominal	Low	High
InitActive (d)	Initial number of active brigades in the Total Force.	18	0	60
InitEnhanced (d)	Initial number of enhanced ready brigades (ERBs) in the Total Force.	15	0	60
InitGuard (d)	Initial number of National Guard brigades in the Total Force.	24	0	60
InitAdvTrainCap (d)	The initial number of CTCs available.	1	1	6
AdvTrainDuration (c)	The amount of time (in months) it takes to complete advanced level training (through division level) for a brigade.	2	1	6
IndTrainCap (d)	The number of brigades that can be trained simultaneously for armor basic and advanced MOS training.	4	1	10
IndTrainDuration (c)	The amount of time (in months) to train recruits through armor basic and advanced MOS training.	4	2	12
SiteExpandDelay (c)	The amount of time (in months) before the maximum number of CTCs are available.	3	1	12
RetrainTime (c)	The maximum amount of time allowed before a unit must be retrained through a CTC.	100	24	100
TrainerReq (c)	Number of heavy brigades required to provide trainers for one CTC.	1.5	1.5	3
InitEqp (c)	Initial number of brigade sets of equipment.	78	20	120
InitProdRate (c)	Initial production rate for the tracked combat vehicle industrial base (brigades per month). The high value assumes the industrial base is hot (1 plant is producing new vehicles using 1 shift).	0	0	0.238
MaxProdRate (c)	The maximum production rate for the tracked combat vehicle industrial base (brigades per month). The nominal rate is based on 1 plant using 2.5 shifts. The high value is based on 3 plants using 2.5 shifts.	0.596	0.238	1.79

Table B.1—continued

Lagtime (c)	The amount of lead time that elapses before production of new vehicles begins.	24	0	60
Ramptime (c)	The amount of time (in addition to lagtime) it takes to ramp up production to MaxProdRate.	12	0	24
CTCMax (d)	Maximum number of CTCs.	6	1	12
IndTrainerReq (c)	The number of heavy brigades required to provide trainers for individual MOS training (provided by the Armor School) .	3	1	6

NOTE: The designators (d) and (c) distinguish between discrete and continuous variables.

HEAVY FORCE EXPANSION TIMING AND COST

In addition to the parameter values for the heavy force expansion timing model, the costing explorations required costing numbers for four of the parameters. They are shown in Table B.2.

Table B.2

Heavy Force Cost Parameters

Parameter	Parameter Description	Nominal	Low	High
InitAdvancedTrainCap FixedCost (c)	The fixed cost for building a CTC.	1,100	400	2,200
InitActiveRecurCost (c)	The annual operating cost of an active brigade.	405	100	1,000
InitEnhancedRecurCost (c)	The annual operating cost of an ERB brigade (estimated from ARNG data).	100	40	160
InitGuardRecurCost (c)	The annual operating cost of an ARNG brigade.	82	40	160
InitAdvTrCapRCost (c)	The annual operating cost of a CTC, including TRADOC and FORSCOM costs for ten rotations a year.	218	100	400

NOTES: All costs are in millions (FY98) for a single brigade within the division, except for InitAdvTrCapRCost, which is FY97.

The designator (c) indicates a continuous variable.

To derive fixed and recurring cost estimates for various types of units, we relied upon a model provided by the Army's Cost and Economic Analysis Center (CEAC).

CEAC provided us the latest version of the Force and Organization Cost Estimating System (FORCES) model.[1] FORCES is a cost model that incorporates detailed database information on all units in the current force structure, both active and reserve.[2] The FORCES model is capable of estimating the cost of different events in a unit's life cycle, including acquisition of resource costs and annual operating costs. FORCES contains detailed data (including all subordinate units in the division) on the following units:

Major Unit	Component
101st Air Assault Division	AC
10th Infantry Division (Light)	AC
1st Armor Division	AC
1st Cavalry Division	AC
1st Infantry Division (Mech)	AC
25th Infantry Division (Light)	AC
28th Infantry Division	ARNG
29th Infantry Division (Light)	ARNG
2nd Infantry Division	AC
34th Infantry Division	ARNG
35th Infantry Division (Mech)	ARNG
38th Infantry Division	ARNG
3rd Infantry Division (Mech)	AC
40th Infantry Division (Mech)	ARNG
42nd Infantry Division	ARNG
49th Armor Division	ARNG
4th Infantry Division (Mech)	AC
82nd Airborne Division	AC

We used the CEAC FORCES model to estimate the fixed cost of a division by using the FORCES Acquisition of Resources cost estimate.

---

[1]We are grateful to Mr. George Michaels and Mr. Jean Duval for providing FORCES to us.

[2]FORCES tracks TOE units by Standard Requirements Codes (SRC). FORCES does not differentiate between guard (ARNG) and enhanced ready brigades (ERBs).

The recurring cost of a division was computed as the FORCES Annual Operations cost estimate.[3] Since our exploratory model only requires reasonable cost estimates to begin with, we chose representative divisions from the current force structure to derive "standard" fixed and recurring costs within FORCES. Since we did not want to ignore any division-level assets (such as aviation and logistical units) that the FORCES model includes in its calculations, we computed the total cost for each division and divided by three to get a rough estimate of what a brigade costs. The representative units used to derive standard recurring and fixed costs are listed below:

Unit Type	Representative Unit	Fixed Cost	Recurring Cost
Heavy active	1st Armored Division	$1,447	$405
Heavy ARNG	35th Mechanized Division	$913	$82

All costs are in millions (FY98) for a single brigade within the division.

Fixed and Recurring Costs for Combat Training Centers (CTC)

For the expansion of heavy units, we used the National Training Center (NTC) at Fort Irwin as the model CTC. There are three major recurring costs for running the NTC:[4]

- OPFOR and Opsgroup—all the TRADOC costs of running the ranges, providing the opposition force (the 11th ACR), trainers, observers, controllers, etc.

---

[3]Acquisition of Resources is defined as the cost to procure the materiel and personnel required for the selected unit. It includes the costs of training personnel through initial MOS training. Annual Operating costs is defined as the direct and indirect costs to operate the selected force at the specified readiness and optempo levels. For a detailed description and breakdown of these cost estimates, refer to *The FORCES Cost Model (FORCES 97.1), Introduction*, June 1997, Department of the Army, CEAC.

[4]The Directorate for Resource Management (DRM) at Fort Irwin accounts for all the costs associated with the NTC, including the total costs for the 11th ACR, the prepo fleet, and the G3 Office (Plans and Operations for the Post).

- FORSCOM deployments—all the transportation costs for the BLUEFOR units that must transport personnel and equipment to the NTC.

- "Infrastructure" costs for the NTC.[5]

After consulting with numerous analysts, we estimate the NTC annual operating cost to be roughly between $203 million and $218 million at current optempo (10 rotations a year).[6] We assume this is a reasonable estimate for the recurring cost of a generic CTC built along the lines of the NTC.

The fixed cost of the NTC was estimated indirectly by computing the plant replacement value (PRV) for the NTC. The PRV is the current (FY96) cost of replacing or replicating the NTC. The PRV is $1.103 billion in FY96 dollars.[7]

LIGHT FORCE EXPANSION TIMING

Table B.3 gives the names, descriptions, and values for the parameters in the model used to explore the timing of light force expansion. These values were derived from the research documented in Chapter Six.

---

[5]Costs associated with the NTC at Fort Irwin are hard to pin down because many of these costs are "outside the training box" but still support the NTC. Some logistics support accounts are currently not costed under NTC mission support.

[6]Including Cathy Zimmerman, Elizabeth Walker, and Dwight Schalles at Directorate of Resource Management, Fort Irwin, CA; Wendy Freeman, NTC Program Analyst, CTC Division, TRADOC, Fort Monroe, VA.

[7]Provided by Lora Muchmore, a budget analyst at DUSD (Industrial Affairs and Installations), Office of Analysis and Investment Directorate.

Table B.3

Light Force Timing Parameters

Parameter	Parameter Description	Nominal	Low	High
InitActive (d)	Initial number of active brigades in the Total Force	12	0	24
InitEnhanced (d)	Initial number of enhanced ready brigades (ERBs) in the Total Force	15	0	60
InitGuard (d)	Initial number of National Guard brigades in the Total Force	24	0	60
InitAdvTrainCap (d)	The initial number of JRTCs available	1	1	6
AdvTrainDuration (c)	JRTC training time (months)	0.5	1	2
NGTrainDur (c)	Training time necessary before an ARNG brigade is ready for a JRTC (months)	2	0.5	12
IndTrainCap (d)	The number of individual training sites for MOS training	4	1	20
IndTrainDuration (c)	The amount of time to complete basic and advanced MOS training (months)	4	2	12
SiteExpandDelay (c)	Time before advanced training sites (JRTCs) are available (months)	6	3	12
RetrainTime (c)	Time before a unit must be retrained through a JRTC (months)	100	24	100
TrainerReq (c)	Number of ready light brigades to provide trainers for one JRTC (brigades)	1.5	1.5	3
CTCMax (d)	Maximum number of JRTCs	1	1	12
IndTrainerReq* (c)	The number of light brigades required to provide trainers for individual MOS training (provided by the Infantry School)	1.5	1	3

NOTE: The designators (d) and (c) distinguish between discrete and continuous variables.

*Both IndTrainerReq and TrainerReq were determined by comparing the number of critical trainers needed—based on either the TDA of the 29th Infantry Regiment or of the cadre and OPFOR at the JRTC—with the number of critical soldiers of particular rank/MOS that could be supplied from a brigade (using the TOE of a standard light division).

LIGHT FORCE EXPANSION TIMING AND COST

For the training of light units, the Joint Readiness Training Center (JRTC) was used as the model CTC. The light force cost parameters are shown in Table B.4.

Table B.4

Light Force Cost Parameters

Parameter	Parameter Description	Nominal	Low	High
InitAdvancedTrainCap FixedCost (c)	Fixed cost for a JRTC	1,700	400	3,200
InitActiveRecur Cost (c)	Annual operating cost of an active brigade	234	100	1,000
InitEnhancedRecurCost (c)	Annual operating cost of an ERB brigade (assumed to be the same as the ARNG)	60	46	185
InitGuardRecur Cost (c)	Annual operating cost of an ARNG brigade	52	40	160
InitAdvTrCapRCost (c)	Annual operating cost of a JRTC, including TRADOC and FORSCOM costs for ten rotations a year	110	50	400

NOTES: All costs are in millions (FY98) for a single brigade within the division except for InitAdvTrCapRCost, which is FY97.

The designator (c) indicates a continuous variable.

As with the heavy force cost estimates, we used the CEAC FORCES model to estimate the fixed cost of a division by using the FORCES Acquisition of Resources cost estimate. The representative units used to derive standard recurring and fixed costs are listed below:

Unit Type	Representative Unit	Fixed Cost	Recurring Cost
Light Active	25th Light Infantry Division	$517	$234
Light ARNG	29th Light Infantry Division	$441	$53

NOTE: All costs are in millions (FY98) for a single brigade within the division.

Bankes, Steven, "Exploratory Modeling for Policy Analysis," *Operations Research*, Vol. 41, No. 3, 1993, pp. 435–449.

Bankes, Steven C., "Computational Experiments and Exploratory Modeling," *CHANCE*, Vol. 7, No. 1, 1994, pp. 50–57.

Bankes, Steven C., and J. J. Gillogly, *Exploratory Modeling: Search Through Spaces of Computational Experiments*, Santa Monica, CA: RAND, RP-345, 1994.

Brinkerhoff, John R., "Reconstitution: A Critical Pillar of the New National Security Strategy," *Strategic Review*, Vol. 29, No. 4, Fall 1991, pp. 9–22.

Brooks, Arthur, S. C. Bankes, and B. E. Bennett, *Weapon Mix and Exploratory Analysis: A Case Study*, Santa Monica, CA: RAND, DB-216/2-AF, 1997.

CBO Papers, *Alternatives for the U.S. Tank Industrial Base*, Washington, D.C.: Congressional Budget Office, February 1993.

Industrial Assessment for Tracked Combat Vehicles, Office of the Under Secretary of Defense, Industrial Capabilities and Assessments, October 1995.

Goldich, Robert L., *Defense Reconstitution: Strategic Context and Implementation*, CRS Report for Congress, November 20, 1992.

Kreidberg, LTC Marvin A., and LT M. Henry, *History of Military Mobilization in the U.S. Army 1775–1945*, Department of the Army

Pamphlet No. 20-212, Washington, D.C.: Department of the Army, June 1955.

Lempert, R. J., M. E. Schlesinger, and S. C. Bankes, "When We Don't Know the Costs or the Benefits: Adaptive Strategies for Abating Climate Change," *Climatic Change*, No. 33, 1996.

Lippiatt, Thomas F., J. C. Crowley, P. K. Dey, and J. M. Sollinger, *Postmobilization Training Resource Requirements*, Santa Monica, CA: RAND, MR-662-A, 1996.

Lippiatt, Thomas F., J. C. Crowley, and J. M. Sollinger, *Time and Resources Required for Postmobilization Training of AC/ARNG Integrated Heavy Divisions*, Santa Monica, CA: RAND, MR-910-A, 1998.

Palmer, Robert R., Bell I. Wiley, and William R. Keast, *United States Army in World War II, The Army Ground Forces: The Procurement and Training of Ground Combat Troops*, Department of the Army: Washington, D.C., 1948.

Peters, Ralph, "Our Soldiers, Their Cities," *Parameters*, Vol. 26, No. 1, Spring 1996, pp. 43–50.

Schubert, Frank N., *Mobilization: The U.S. Army in World War II, The 50th Anniversary*, CMH Pub 72-32 (no date).

Structuring U.S. Forces After the Cold War: Costs and Effects of Increased Reliance on the Reserve, Congressional Budget Office, September 1992.

TRADOC Mobilization and Operations Planning and Execution System 1-97, Annex T, Fort Monroe, VA: Headquarters, U.S. Army Training and Doctrine Command, May 30, 1997.

Transforming Defense: National Security in the 21st Century, Report of the National Defense Panel, December 1997.